Engineering Energy Aluminum Conductor Composite Core (ACCC) and Its Application

Elsevier and CEPP Energy Series

Engineering Energy Aluminum Conductor Composite Core (ACCC) and Its Application

YUAN CHEN
Group Leader and Professor, North China Electric Power Research Institute Co., Ltd, Beijing, China

中国电力出版社
CHINA ELECTRIC POWER PRESS

ELSEVIER

ACADEMIC PRESS
An imprint of Elsevier

Academic Press is an imprint of Elsevier
125 London Wall, London EC2Y 5AS, United Kingdom
525 B Street, Suite 1650, San Diego, CA 92101, United States
50 Hampshire Street, 5th Floor, Cambridge, MA 02139, United States
The Boulevard, Langford Lane, Kidlington, Oxford OX5 1GB, United Kingdom

British Library Cataloguing-in-Publication Data
A catalogue record for this book is available from the British Library

Library of Congress Cataloging-in-Publication Data
A catalog record for this book is available from the Library of Congress

ISBN: 978-0-12-815611-7

Publisher: Joe Hayton
Acquisition Editor: Lisa Reading
Editorial Project Manager: John Leonard
Production Project Manager: Anitha Sivaraj
Cover Designer: Christian J. Bilbow

Typeset by MPS Limited, Chennai, India

CONTENTS

PREFACE

Carbon fiber composite core aluminum conductor (ACCC) has a lot of advantages, such as light weight, low coefficient of linear expansion, high strength, high elastic modulus, high temperature resistance, fatigue resistance, corrosion resistance and, other technical advantages. It can not only be used to improve the power transmission capacity, but also effectively improve the safe operation of the line and reduce the transmission loss, and so is one of the new wires with potential for development. This book comprises a comprehensive discussion of carbon fiber ACCC technology and its development based on the domestic independent research and development results in China. This book consists of eight chapters, including a summary of the new wire technology; carbon fiber ACCC manufacturing; carbon fiber ACCC testing; carbon fiber ACCC supporting fittings; construction, operation, and maintenance; carbon fiber ACCC stress—sag calculation, carbon fiber ACCC standardization, scope of application for carbon fiber ACCC, and carbon fiber ACCC applications.

This book is both intended for those people who are engaged in fiber composite core conductor research and manufacturing, as well as transmission conductor design, construction, operation, and maintenance personnel, and as a self-study book and reference book for professional teachers and students in related tertiary institutions.

INTRODUCTION

Carbon fiber composite core aluminum conductor has the advantages of light weight, low linear expansion coefficient, high strength, high elastic modulus, high temperature resistance, fatigue resistance, corrosion resistance, and other technical advantages. It is one of the new conductor varieties with potential for development. In 2006, State Grid Corporation of China (SGCC) included the product in the first batch of key applications in its new technology directory. In 2008, it also started independent research into and application of the product. In 2010, the project was included in the National High-tech Development Research Program (863 Program), which promoted the rapid development of carbon fiber composite core aluminum conductor technology. Conductors, ancillary products, design, construction, operation and maintenance, and technical standards are increasingly mature. With the improvement of domestic technology and the product prices gradually returning to normal, carbon fiber composite core aluminum conductors can play an important role in the construction and upgrade of power grids.

This book is hereby written to provide the power system and related industries a more extensive and comprehensive understanding and mastery of this technology, so that carbon fiber composite core aluminum conductor can play its due role.

This book is a comprehensive review and summary of carbon fiber composite core aluminum conductor technology, not only including the basic principles of products, but also including deeper product manufacturing, supporting technology, etc. It can be used as important reference material for carbon fiber composite core aluminum conductor research, manufacturing and manufacturing personnel, and transmission conductor design, construction, operation, and maintenance personnel.

I received a lot of help from my colleagues and other power industry and manufacturing industry personnel during this writing, so I sincerely thank everyone involved.

Yuan Chen
December 2018

LIST OF CO-AUTHORS

Yuan Chen
North China Electric Power Research Institute Co., Ltd

Lu Yi
North China Electric Power Research Institute Co., Ltd

Gao Shun'an
North China Electric Power Research Institute Co., Ltd

Zhang Xu
North China Electric Power Research Institute Co., Ltd

Wang Xin
North China Electric Power Research Institute Co., Ltd

Fan Shuochao
North China Electric Power Research Institute Co., Ltd

Gao Yanfeng
North China Electric Power Research Institute Co., Ltd

Song Furu
He Bei Silicon Valley Chemical Co., Ltd

Zhou Guohua
State Grid Shanxi Electric Power Company

Xue Wenxiang
State Grid Jibei Electric Power Co., Ltd

Wang Shuyuan
State Grid Jibei Electric Power Co., Ltd

Zhang Jifei
State Grid Jibei Electric Power Co., Ltd

Wang Hui
State Grid Jibei Electric Power Co., Ltd

CHAPTER 1

New Conductor Technology Summary

Contents

With the drastic increase of power load and the increasingly difficult choice of transmission corridors, the demand for improving the transmission power of corridors has become more and more urgent. The regular aluminum conductor steel reinforced (ACSR) can hardly meet the transmission demand. Given this background, a series of conductors that can be used for capacity expansion of transmission are developed both at home and abroad, including large-section ACSR and thermal-resistant aluminum alloy conductor (TACSR), zirconium alloy aluminum conductor invar steel reinforced, aluminum-based ceramic fibers, core heat-resistant aluminum alloy conductors, carbon fiber aluminum conductor composite core (ACCC), and other new conductors. Large-section ACSR is still part of the scope of ACSR, achieving the purpose of capacity expansion by just increasing the aluminum cross-section. Steel core heat-resistant aluminum alloy wire (TACSR) is made of aluminum added with zirconium, so as to improve the heat resistance of aluminum, with a

Engineering Energy Aluminum Conductor Composite Core (ACCC) and its Application
DOI: https://doi.org/10.1016/B978-0-12-815611-7.00001-6

wide operating temperature range. Zirconium alloy aluminum conductor invar steel reinforced is made of steel added with the appropriate proportion of nickel, so that the core wire has a lower linear expansion coefficient in a wide operating temperature range. Aluminum-based ceramic fiber core aluminum conductor (aluminum conductor composite reinforced (ACCR)) and carbon fiber ACCC, collectively referred to as composite core aluminum conductors, are the conductors with the best overall performance at present. They not only have a wide operating temperature range and lower linear expansion coefficient, but also have a lower unit weight. ACCCs break the traditional conductor structure, which has many technical advantages, and are an important innovation in transmission conductors.

1.1 NEW CONDUCTOR TECHNOLOGY

1.1.1 Comparison of Production Processes

1. Large-section ACSR

   Large-section ACSR is a just an ordinary conductor with a large aluminum cross-section and higher flow capacity. In essence, it does not belong to the category of new conductors. The unit weight and outer diameter of the large-section ACSR used for the capacity expansion of reconstruction are larger than the original conductor. Therefore, the tension and sag of the conductor under various operating conditions exceed that of the original conductor correspondingly. The strength, height, and other parameters of the tower are matched with the original conductor, and the large cross-section conductor inevitably needs to replace the tower (including increased strength and increased height), the result being that the amount of line reconstruction workload and capital investment is too large.

2. TACSR

The structure and production process of TACSR is basically the same as with ACSR, and is shown in Fig. 1.1. The main difference is that conventional hard aluminum wire is used instead of heat-resistant aluminum alloy wire.

The production process of heat-resistant aluminum alloy wire includes solid solution type and solid solution precipitation type. The addition of metallic zirconium element to the aluminum material to form tiny Al_3Zr prevents recrystallization of the metal, thereby increasing the heat resistance of the aluminum material; the loss of the strength won't go over

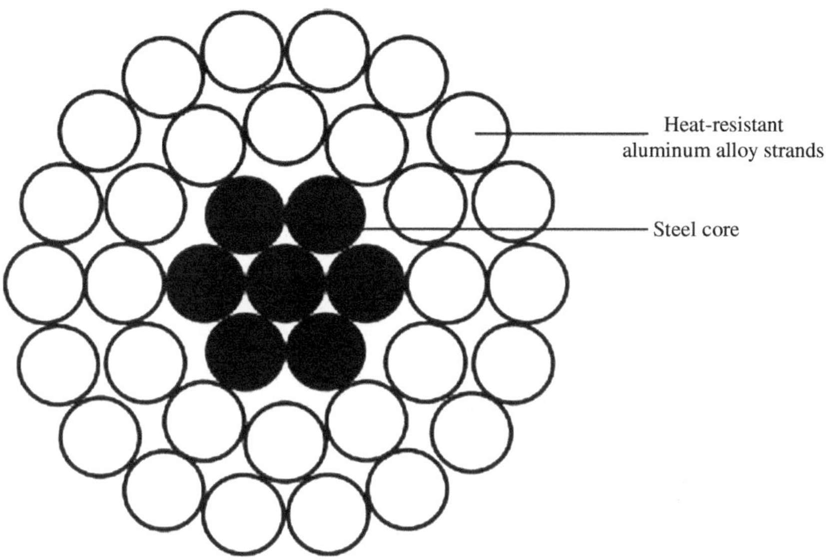

Figure 1.1 Structure diagram of TACSR.

the allowable value when operating at a higher temperature. However, the addition of zirconium has a detrimental effect on the electrical conductivity, that is, the heat resistance of heat-resistant aluminum alloy strands and the electrical conductivity of the heat-resistant aluminum alloy constitute a pair of contradiction.

Heat-resistant aluminum single wire has four types: AT1, AT2, AT3, and AT4; see the correspondence model of the series in Table 1.1. The conductivity of 58% international annealed copper standard (IACS) ordinary heat-resistant aluminum alloy single and super heat-resistant aluminum alloy single wire have a higher resistivity and line loss is large; it has been replaced by a single wire with a higher conductivity (60% IACS[1]) heat-resistant aluminum alloy.

The tensile strength of the heat-resistant aluminum alloy single conductor is basically the same as that of the hard aluminum conductor at room temperature. Therefore, the mechanical properties of the ordinary conductor and the heat-resistant aluminum alloy stranded conductor are basically the same when they have the same cross-section. The mechanical properties have no essential difference between it and the ordinary conductor in the circuit design.

[1] International annealed copper standards, the conductivity of 58 ms/m is identified as per the 100% IACS.

Table 1.1 Correspondence model table of heat-resistant aluminum alloy steel core

Wire type	Japanese model	IEC model	Conductivity (%IACS)	Tensile strength (MPa)	Long-term allowable temperature (°C)
58% heat-resistant aluminum alloy	58TA1		58	159–169	150
60% heat-resistant aluminum alloy	60TA1	AT1	60	159–169	150
High-strength heat-resistant aluminum alloy	KTA1	AT2	55	225–248	150
Super heat-resistant aluminum alloy	UTAI		58	159–169	200
High conductive super heat-resistant aluminum alloy	ZTA1	AT3	60	159–176	210
Special heat-resistant aluminum alloy	XTA1	AT4	58	159–169	230

3. Zirconium alloy aluminum conductor invar steel reinforced

   Zirconium alloy aluminum conductor invar steel reinforced structure diagram is shown in Fig. 1.2. It is made by replacing TACSR; heat-resistant aluminum alloy wire is a trapezoidal wire. Zirconium alloy aluminum conductor invar steel reinforced was developed in the early 1980s. Invar is a nickel–iron alloy consisting of 36% nickel, 63.8% iron, and 0.2% carbon. This conductor has a lower linear expansion coefficient over a wide temperature range.

   One production process of invar alloy is the usage of vacuum induction furnace or non–vacuum induction furnace. After thermal

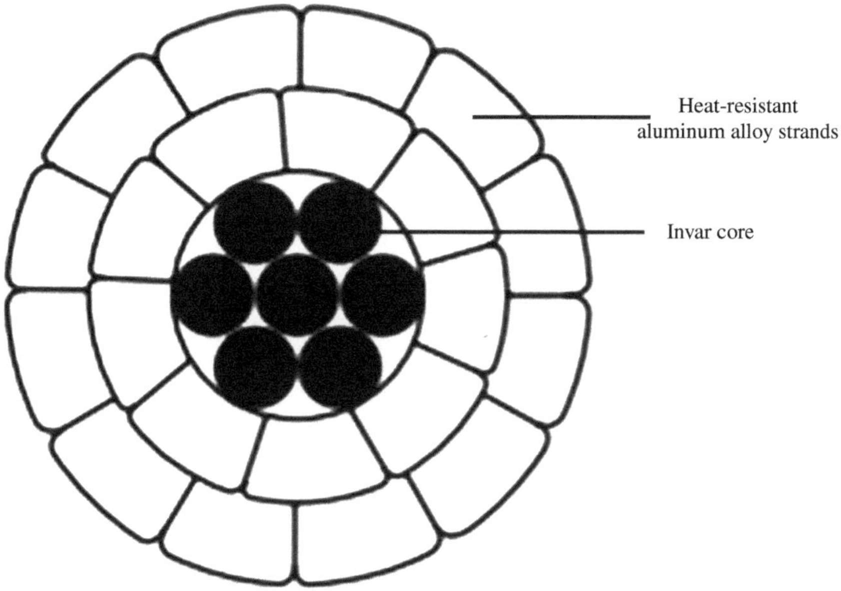

Figure 1.2 Structure diagram of zirconium alloy aluminum conductor invar steel reinforced.

processing, mild cold deformation, and pickling, it is then then cold-processed into wire. Invar alloy wire can be further made into galvanized invar alloy wire or aluminum clad invar alloy wire, and then twisted with heat-resistant aluminum alloy wire, and made into zirconium alloy aluminum conductor invar steel reinforced.

4. Aluminum-based ceramic fiber core aluminum wire

The structure of aluminum-based ceramic fiber core aluminum wire is basically the same as that of ACSR (see Fig. 1.3). The difference is that the material or the shape of the aluminum wire is different. The steel core is replaced by the aluminum-based ceramic fiber core, heat-resistant aluminum alloy wire replaces the hard aluminum, and heat-resistant aluminum alloy line can be either circular or trapezoidal in structure. Each core of the aluminum-based ceramic fiber is composed of thousands of 12-μm-thick high-strength aluminum oxide fibers and ceramic fibers embedded in the aluminum core wire continuously in the same direction.

The production process of aluminum-based ceramic fiber core aluminum wire core is as follows. A crushing device is used to crush and remove slag for the fibers, the vacuum pressure filtration method is used to make the trapezoidal preformed unit, and then the trapezoidal composite material is manufactured by the extrusion—infiltration casting method.

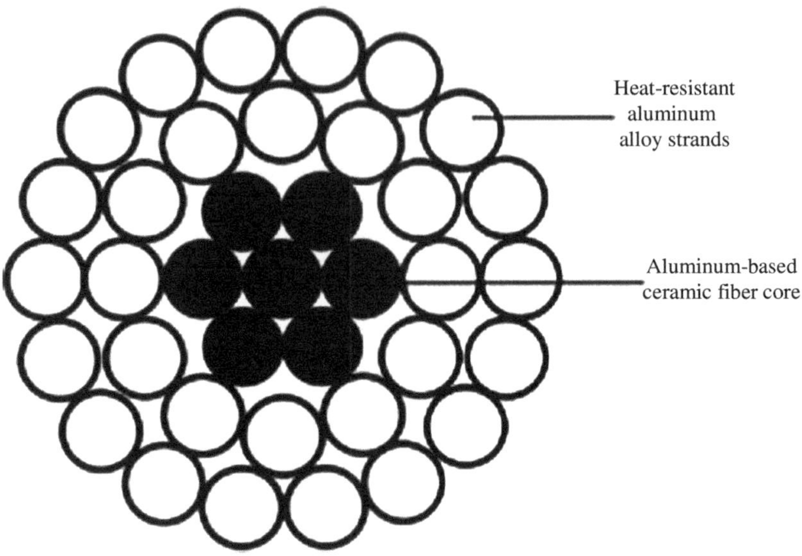

Figure 1.3 Aluminum-based ceramic fiber core aluminum stranded wire structure diagram.

In this process, the well-preheated preformed unit is first placed into the mold, pouring refined aluminum alloy liquid and pressuring. Under pressure, aluminum liquid penetrates into the gaps of the preformed units. The pressure is retained until the aluminum alloy liquid is fully solidified; after the pressure is removed, the composite material is obtained. By adjusting and squeezing the preheating temperature and squeeze casting pressure, speed, and other parameters of preformed unit and casting mold to ensure complete penetration of prefabricated parts a high-quality gradient composite material is obtained. Composite core wire and heat-resistant aluminum alloy wire is stranded into the aluminum-based ceramic fiber core aluminum wire.

5. Carbon fiber aluminum conductor composite core

Carbon fiber aluminum conductor composite core comes from the United States, Japan, and other developed countries, and the products from US CTC include the carbon fiber composite core soft aluminum line structure; see Fig. 1.4. The difference between the carbon fiber aluminum conductor composite core and the traditional ACSR is the material and the shape of the core wire and the aluminum wire. The single high-strength heat-resistant carbon fiber composite core replaces the multistrand steel core, and the main raw material of the carbon fiber composite core is carbon fiber and high-temperature curing resin, which is

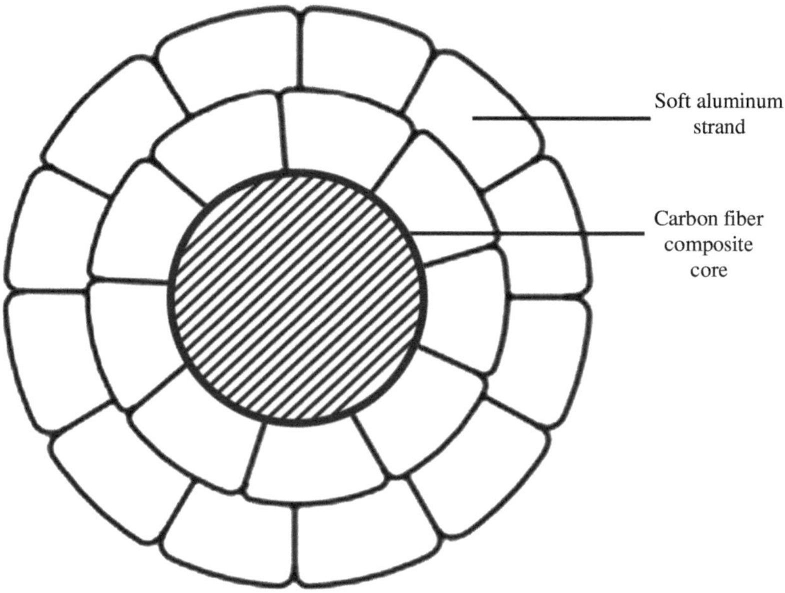

Figure 1.4 Carbon fiber composite core structure diagram of soft aluminum wire.

carbonized by the polyacrylonitrile (PAN) formed after the pultrusion process. US CTC's products replace round hard aluminum with high-conductivity soft aluminum trapezoidal aluminum, and carbon fiber composite core and soft aluminum trapezoidal aluminum become the carbon fiber aluminum conductor composite core.

However, because of the low mechanical strength and low hardness of the soft aluminum, which is easily damaged during construction and overhaul, the construction process is relatively complicated. Therefore, the carbon fiber composite core aluminum conductor independently developed by State Grid Corporation of China has also used a heat-resistant aluminum alloy conductor in recent years to solve the low mechanical strength problems of aluminum alloy, and trial production of hard-aluminum strand carbon fiber composite core aluminum conductor is underway to take into account both the electrical conductivity of aluminum strands and mechanical strength.

1.1.2 Performance Comparison

1. Core density and strength comparison

Table 1.2 compares the physical and mechanical properties of the core conductors of several types of conductors. From the data in the table, the

Table 1.2 Physical and mechanical properties of several conductor cores

Conductor type	ACSR	TACSR	Zirconium alloy aluminum conductor invar steel reinforced	ACCR	ACCC
Core material	Galvanized (aluminum cladding) steel wire	Galvanized (aluminum cladding) steel wire	Galvanized (aluminum cladding) invar steel wire	Aluminum-based ceramic fiber core	Carbon fiber composite core
Density (g/cm^3)	7.8	7.8	7.1	3.3	1.6–1.9
Tensile strength (MPa)	1300	1300	1080	1275	2400
Elastic modulus (GPa)	200	200	152	216	120

Table 1.3 Several expansion conductor heat resistance comparisons

Conductor type	Allow continuous use temperature (°C)	Migration point temperature (°C)	Migration point below the linear expansion coefficient (°C^{-1})	Migration point above the line of expansion Number (°C^{-1})
ACSR	≤90	120	19.4×10^{-6}	11.5×10^{-6}
TACSR	≤150	120	19.4×10^{-6}	11.5×10^{-6}
Zirconium alloy aluminum conductor invar steel reinforced	≤200–210	100	17×10^{-6}	3.7×10^{-6}
ACCR	≤210	130	16.3×10^{-6}	6.3×10^{-6}
ACCC	≤160	85	$12.8–19.6 \times 10^{-6}$	2.0×10^{-6}

density of the carbon fiber composite core is the smallest, only 1/2 of that of the aluminum-based ceramic fiber core, which is 1/4 of that of the other core conductors. Tensile strength of carbon fiber composite core is the maximum, about twice that of the steel core, invar core, and aluminum-based ceramic fiber core. Specific strength of carbon fiber composite core aluminum conductor is the largest, aluminum-based ceramic fiber core conductor is second, and the effect of weight of the conductor for the arc sag is the most significant, so at the same initial temperature and conductor tension conditions, carbon fiber composite core aluminum conductor has the smallest sag.

2. Comparison of heat resistance

Several expansion conductor thermal performance comparison in Table 1.3.

As can be seen from Table 1.3, carbon fiber composite core aluminum conductor and other expansion conductor allows continuous use temperature above 150°C, much higher than ACSR's 90°C, and has a smaller linear expansion coefficient.

When they have the same current carrying capacity and outside diameter, the sag of carbon fiber composite core soft aluminum conductor is only 1/6 that of ACSR. Figs. 1.5 and 1.6 are temperature—sag and temperature—tension curves for both sizes of 1020 kcmil ACCC/TW conductor and 795 kcmil ACSR. The arc sag characteristics of both are shown in Table 1.4 (test conditions: lead span of 65.5 m, initial tension of lead conductor of 25% RTS).

Under high-temperature conditions, the arc of carbon fiber composite core aluminum conductor wire is small because the linear expansion coefficient above the temperature of the migration point is the smallest, followed by the core steel wire of the invar. Therefore, the difference of

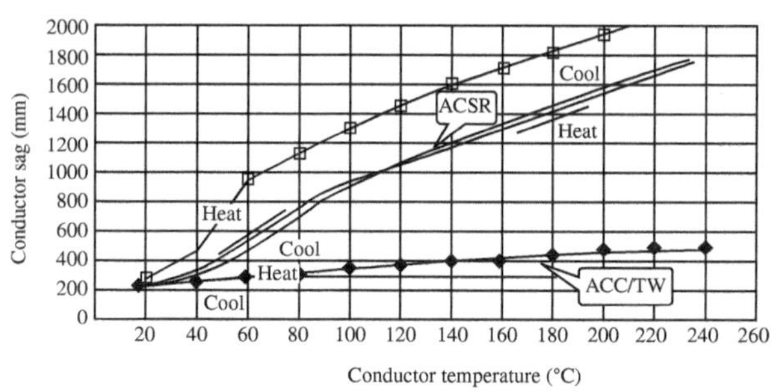

Figure 1.5 ACSR and ACCC/TW conductor temperature—sag curve.[2]

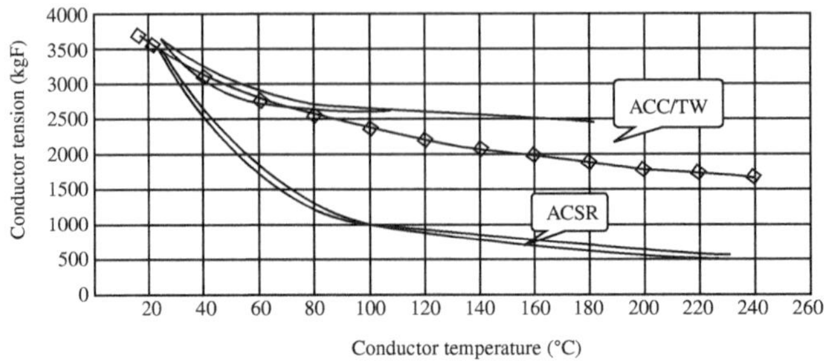

Figure 1.6 ACSR and ACCC/TW conductor temperature–tension curve.

Table 1.4 Comparison of the sag characteristics of ACCC/TW and ACSR

Conductor types and specifications	Sag (m)			Tension (kgF)		
	26°C	183°C	Variable quantity	26°C	183°C	Variable quantity
ACSR (795 kcmil)	236	1422	+ 1186	3566	622	− 2944
ACCC/TW	198	312	+ 144	3745	2471	− 1274

sag between the carbon fiber composite core aluminum conductor and other wires will further increase as the temperature rises.

It can be seen from Table 1.4, under the same experimental conditions, that the sag of the ACSR increases from 236 to 1422 mm, with an increase of 1186 mm with an increase of five times when the temperature is raised from 26 to 183°C. However, the sag of ACCC/TW increases from 198 to 312 mm, which increases by 114 mm, with just an increase of 0.57 times, that is to say, the change of ACCC/TW sag is only 1/10 of ACSR. It can also be seen from Fig. 1.5 that the sag change of ACCC/TW above the inflection point (migration point) is very small, while the ACSR varies considerably. Fig. 1.6 represents the curve of the conductor tension at the corresponding temperature. When the temperature rises from 26 to 183°C, ACSR tension reduces from 3566 kgF (1 kgF ≈ 9.8 N) to 622 kgF, decreased by 82.56%, while the ACCC/TW tension decreases from 3745 to 2471 kgF, decreased by 34.02%. Under 26°C the tension of two conductors is basically the same, whereas at 183°C the tension of the ACSR is only 1/4 of the ACCC/TW.

Due to the lightweight and small coefficient of linear expansion of the carbon fiber aluminum conductor composite core, the altitude of the towers can be lowered, the number of towers can be reduced, and the cost of iron tower foundation can also be reduced.

Table 1.5 Comparison of properties of aluminum conductor for several kinds of expansion conductors

Conductor type	**ACSR**	**TACSR**	**Zirconium alloy aluminum conductor invar steel reinforced**	**ACCR**	**ACCC**
Aluminum strand material	Duralumin	Aluminum alloy	Aluminum alloy	Aluminum alloy	Soft aluminum
Conductivity (20°C)	61% IACS	60% IACS	60% IACS	60% IACS	62% IACS
Aluminum strand cross-section (mm^2)	403	403	413	403	517
Strength of aluminum strand (MPa)	158–183	158–183	158–183	139	60
Short-term allowable temperature (°C)	≤ 120	≤ 180	≤ 240	≤ 240	≤ 160

3. Performance comparison of aluminum conductor

Table 1.5 is the performance comparison of several kinds of expanded conductor aluminum conductors. The electric conductivity is over 62% for carbon fiber composite core aluminum conductor IACS at 20°C, and the transmission loss is 5% lower than that of zirconium alloy aluminum conductor invar steel reinforced and aluminum-based ceramic fiber core aluminum strands. However, the tensile strength of soft aluminum strands with carbon fiber composite core aluminum conductor is only 59 MPa, less than 50% of hard aluminum and aluminum alloy strands.

4. Other performance comparisons

For the core, except for carbon fiber aluminum conductor composite core and aluminum-based ceramic fiber core aluminum wire, the heat-resistant aluminum alloy strands, zirconium alloy aluminum conductor invar steel reinforced, and invar ordinary ACSR are magnetic metal materials, and eddy current magnetic loss and electrochemical corrosion between the core and the aluminum conductor occur in the line running process. However, the carbon fiber composite core, because of the insulation resin package, does not have the eddy current magnetic loss and the electrochemical corrosion between the core and the aluminum conductor, and the environmental corrosion resistance is excellent. In addition, the core line of carbon fiber aluminum conductor composite core has a self-damping structure with better anti–wind vibration ability.

1.2 CARBON FIBER COMPOSITE CORE ALUMINUM CONDUCTOR TECHNOLOGY

1 Technical advantages of carbon fiber composite core aluminum conductor

Overhead transmission lines are generally made of ACSR, with a steel strand in the center, and are mainly used to withstand conductor tension; the outer part uses an aluminum strand, which is mainly used to transmit current and bear part of the conductor tension. Carbon fiber composite core aluminum conductors replace the carbon steel core with a carbon fiber composite core, and the main technical advantages are as follows:

1. This conductor allows a substantial increase in operating temperature, and transmission capacity is increased significantly, saving the transmission corridor. Carbon fiber yarn has an excellent resistance to high temperature performance, and the strength does not decrease at 2000°C. Selecting the appropriate formula and matrix resin, combined with carbon fiber yarn to make the wire composite core, ensures that the mechanical strength is not significantly reduced for the composite core under high temperature. Therefore, when the carbon fiber composite core aluminum conductor runs at high temperature, the external aluminum conductor only bears the role of transmission current to solve the conventional wire high temperature operation of the mechanical strength of the problem, which can improve the allowable operating temperature wire to improve transmission capacity. Taking LGJ-400/35 ACSR replaced by the carbon fiber composite core aluminum conductor with the same aluminum cross-section as an example, according to the operating temperature of 160°C, the ambient temperature of 40°C, the sunlight intensity of 300 mW/m^2, and the wind speed of 0.5 m/s, the allowable operating temperature of the wire increases from 70 to 160°C, 732 A increases to 1452 A, and transmission power increases by 98%; that is, a line using carbon fiber composite core aluminum conductor can reach the transmission capacity of two conventional lines.
2. The quality of conductor is light, the coefficient of linear expansion is small, and the conductor is small, so the requirements of tower strength, tower height, and tower foundation are reduced.

The strength of carbon fiber composite core is much higher than that of the steel core; under the same tensile force, the quality of carbon fiber composite core is less than 20% of the steel core. The overall quality of the conductor is also significantly reduced; compare the same size LGJ300/50-type steel core aluminum (mass per unit length: 1.21 kg/m, rated tensile force: 103.4 kN) with the JLNR/F2B-300/65 carbon fiber composite core aluminum conductor (mass per unit length: 0.92 kg/m, rated tensile force: 170.5 kN). The weight of the carbon fiber composite core aluminum conductor is only 76% of the ACSR, while the tensile strength is 65% higher than the latter. Therefore, under the premise of considering the elastic modulus, linear expansion coefficient, and creep property of the conductor, if the maximum tension of the conductor and the average running tension of the conductor remain unchanged, the carbon fiber composite core aluminum conductor can replace the conventional conductor, and the sag can be significantly reduced. For example: in 500 kV Wanshun three lines (Wanquan—Shunyi) 471—472 level replaced with carbon fiber composite core aluminum conductor, the sag reduced 3—4 m at the same wire temperature, even considering the carbon fiber composite core aluminum conductor allows the operating temperature of 160°C, far higher than the conventional 70—80°C. But the carbon fiber composite core aluminum conductor, in transfer the temperature higher than the linear expansion coefficient close to zero, the sag increase is limited, can still guarantee the advantage arc sag of the carbon fiber composite core aluminum conductor under the highest operating temperature. Thus, (1) the new line using carbon fiber composite core aluminum conductor may be appropriate to increase the span, reduce the number of poles or tower height, save towers or tower material, or reduce the strength of the tower and tower foundation requirements; and (2) replacing the carbon fiber composite core aluminum conductor with the running line not only eliminates the need to replace the tower or strengthen the tower, but also effectively increases the distance between the conductor and the ground, and the distance between the trees, buildings, and mountains around the line, so as to improve the safe operation margin of the line.

3. Carbon fiber composite core has excellent fatigue resistance, to ensure long-term safe and reliable operation of the line. A large proportion of the mechanical failures of overhead power

transmission lines is related to the fatigue fracture of metal materials. After tens of millions of tensile tests, the residual strength of carbon fiber composite can still reach 90% of the original strength by using suitable resin matrix and formula to make self-damping composite materials with carbon fiber yarn. Under the same test conditions, the residual strength of steel is only 30%–40% of the initial strength.

4. Carbon fiber composite core has excellent anticorrosion and aging-resistant qualities, to ensure long-term safe and reliable operation of the line. With a suitable resin matrix and formulations, composite materials made of carbon fiber yarn have excellent corrosion resistance and resistance to atmospheric aging properties, with no need to maintain under natural conditions. The material has corrosion resistance and antiaging ability for more than 100 years, which is a powerful guarantee for the long-term safe and reliable operation of carbon fiber composite core conductors.
5. Carbon fiber composite core has ultrahigh strength at normal temperature and high temperature. It can bear all the conductor tension in operation, so that the external aluminum strands can use metal materials with lower mechanical strength and better conductivity, and effectively reduce the line loss. The aluminum strands of the carbon fiber composite core aluminum conductor increased the electrical conductivity from 61% IACS to 62% IACS if soft aluminum was used instead of the hard aluminum of conventional wire; and in carbon fiber composite core instead of steel core, composite core will not produce eddy current loss. Therefore, under the same aluminum section cross-section and the same transmission power, the soft aluminum and composite core can effectively reduce the line loss.

2 Problems exist in carbon fiber composite core aluminum conductor

The combination of soft aluminum and carbon fiber composite core can effectively reduce the line loss, but due to the low mechanical strength index of soft aluminum, the composite core wire is easily damaged after being combined with the composite core. There are some disadvantages in the construction and operation, such as deformation and damage of aluminum strands when the outlet aerodyne is used to inspect the conductors. In addition, there are problems such as the carbon fiber composite core aluminum conductor cannot be implemented, continuous tension stringing due to the oversize of wedging fittings designed by foreign technology (including diameter, length), the price of carbon fiber composite core aluminum conductor is relatively high, etc. In recent years the national Power Grid Corp's independent research and development into carbon fiber composite core aluminum conductor has been gradually improving and solving these problems.

1.3 THE BACKGROUND OF INDEPENDENT RESEARCH AND DEVELOPMENT OF CARBON FIBER COMPOSITE CORE ALUMINUM CONDUCTOR

1.3.1 The Need of Power Grid Construction and Upgrading

1.3.1.1 The Need to Improve the Transmission Capacity of the Power Grid

Due to the limited land resources in China, the choice of transmission corridors is restricted and the demand for improving transmission power of corridors in units is increasingly urgent. Speeding up the upgrading of power grids at all levels, improving the transmission capacity and efficiency of the grid, saving resources, and reducing consumption have been listed as the key tasks of the National Power Grid Corp power grid construction.

The power transmission capability depends on the thermal stability of the long distance conductor. The ways to improve the transmission capacity of the above lines are mainly divided into the following two categories.

The first category is not modified on the line ontology, but through the following two ways to improve transmission capacity: (1) by increasing the allowable operating temperature of the conductor, e.g., an ordinary ACSR allows operating temperature to be increased from 70°C to 80–90°C; and (2) by using a conductor that allows the operating temperature to remain the same. However, an online monitoring device that detects conductor temperature and ambient temperature is installed on the wire to obtain line operating data in real time and account for wire carrying capacity. Dynamic management of wire feeding power is carried out to tap the potential of the transmission line.

The second category is through the direct replacement of a large-capacity conductor (such as large-section ACSR, TACSR, etc.) to improve line transmission capacity.

In comparison, the second method is more effective. However, due to the increasing mass of large-capacity conductors (such as large-section ACSR wires), it is necessary to replace or reinforce the tower while replacing the wires. For TACSR, the increase of the sag and the decrease of the distance to the ground (tree, building) must also be considered when the conductor runs at high temperature. Therefore, the improvement of the transmission capacity of transmission lines requires a new type of conductor with superior comprehensive performance, such as carbon fiber aluminum conductor composite core.

1.3.1.2 Need of Improving the Safe Operation Level of the Power Grid

In recent years, along with the frequent occurrence of extremely severe weather such as gales and ice, the wind-induced lead conductor and galloping caused frequent line faults, which seriously affect the safe and stable operation of the power grid. From the end of February to early March 2007, the emergence of a rare snowstorm in bad weather conditions in north China led to the tripping of several 500 kV lines and belowground lines in the power grid, resulting in the shutdown of multiple power plants and the reduction of output. Bad weather including gales and blizzards hit north China again in March, November, and December 2012, with several 500 kV major lines frequently tripped or even disconnected. Mainly due to gale and heavy snow (blizzard) severe weather conditions associated with microtopography and micrometeorological conditions, the swing of conductor (ground wire) covered with ice, windage (including different periods of swing), sleet jump, etc., was led, resulting in distance of the phase-to-phase, phase-to-ground changing, and at the same time the wind resistance and galloping ability of compact transmission lines was very low due to its own structural features, eventually resulting in phase-to-phase, phase-to-ground discharge trip and ground wire damage. In addition, in recent years, construction machinery, vehicle collisions, and other human factors have also caused serious hidden troubles for safe operation of the line. Replacement of lightweight, low-sag, high-strength, fatigue-resistant carbon fiber composite core aluminum conductors can effectively inhibit the above failures and defects to a certain extent to improve the margin of stability operation of grid security.

1.3.1.3 Requirement for Reduction of Power Losses

High-strength composite core can withstand all the tension of carbon fiber composite core aluminum conductor, so the outer layer of carbon fiber composite core can use aluminum strands (such as soft aluminum), which have better conductivity and lower mechanical strength. In addition, there is no eddy current loss in the operation of carbon fiber reinforced composite core, so these factors can effectively reduce transmission losses.

1.3.2 Needs of Breaking the Technical and Price Monopoly

1.3.2.1 Carbon Fiber Technology Monopoly

The kernel of carbon fiber aluminum conductor composite core is high-strength PAN-based carbon fiber whose carbon content is more than 90%. It is the only material whose mechanical strength won't reduce in the environment of 2000°C. It is an essential material for aerospace and military use. The Western countries have imposed a long-term blockade and embargo on high-performance carbon fiber materials, as important military supplies, in China's national defense industry. The research into high-strength carbon fiber in the world started in the 1960s. After the stable period from the 1970s to the 1980s, and the rapid development of the 1990s, the preparation technology and processes have become basically mature in the early 21st century. Japan's Toray, Toho, and Mitsubishi rayon companies rely on the advantages of carbon fiber yarn production and performance and mass production of high-performance carbon fiber, which make Japan a carbon fiber power; its quality and production volume are in a leading position in the world and its raw silk production accounts for 78% of the world's total output. A few developed countries such as Britain, the United States, and France are also major producers of carbon fiber.

Since the 1960s, China has been conducting carbon fiber research. However, due to less investment, imperfect supporting industries, the foreign technology blockade, and other reasons, its research level has lagged far behind advanced countries in the world. The high-performance carbon fiber of structural materials in aircraft, missiles, rockets, aerospace, and model weapons basically rely on import. T700 carbon fiber research was included in the 10th 5-Year Plan by the Ministry of Science and Technology, and the Ministry continues to increase support in the 11th Five-Year plan in order to completely get rid of foreign constraints, and vigorously promote the development of the domestic carbon fiber industry. At present, China's carbon fiber research and development has made a big breakthrough. The successful trial production of T700 high-performance carbon fiber was achieved in the laboratory, and in 2009, the first domestic dry jet wet spinning high-performance carbon fiber precursor production line was built. In recent years, many enterprises have already industrialized production of high-performance carbon fiber.

1.3.2.2 Monopoly of Carbon Fiber Composite Core Aluminum Conductor Technology

There are two foreign manufacturers whose carbon fiber composite core aluminum conductor production technology is relatively advanced, namely the United States CTC Energy (Composite Technology Corporation) and the United States Mercury Cable Energy LLC. The carbon fiber composite core aluminum conductor (model ACCC/TW) made by American CTC Company is a single core form. The center of the carbon fiber yarn is an epoxy resin core bar reinforced by carbon fiber wire, and the outer layer is an isolation layer composed of glass fiber and epoxy resin.

Japan's largest steel wire and cable manufacturer, Japan Tokyo System Co., Ltd., also began to develop carbon fiber reinforced composite core, allowing other wire manufacturers to produce carbon fiber composite core aluminum conductors. The structure of the composite core strand is different from the rod-shaped structure in the United States. It is a stranded structure. According to the data of Tokyo Rope MFG. Co., Ltd., the stranded structure has the characteristics of bending flexibility and twisting flexibility superior to the rod-shaped core, namely the flexibility of the strand structure composite core should be better than that of the rod composite core. In addition, because it is made of a plurality of fine carbon fiber cores, the external force can be theoretically dispersed, improve the stress conditions, and increase the fatigue resistance of the composite core. Moreover, the impact on the entire core is relatively small when a single core is damaged, namely the reliability and stability of composite core with stranding structure is higher than the rod-shaped core, and the structure is also closer to that of the existing ACSR line. However, this structure of the composite core has not yet been applied in the power industry. All the data come from the laboratory. Problems that may arise in construction, operation, etc. still need to be verified.

Before the national Power Grid Corp approved the independent research and development and application of carbon fiber composite core aluminum conductor, only Far East Composite Technology Co., Ltd. was engaged in the production and sales of carbon fiber composite core aluminum conductor. Far East Composite Technology Co., Ltd. is a wholly owned subsidiary of Jiangsu Far East Holding Group, established by the Far East Group Co., Ltd. and US CTC in 2006. Early on, the carbon fiber composite core aluminum conductor that applied to domestic project was directly imported from US CTC; later, the company's carbon fiber reinforced composite core was still dependent on import, and stranding of aluminum strand wire was carried out in China. Even if the domestication of composite core could be realized in the future, composite core materials (carbon fiber, resin matrix) and formulation, technology, and other core technologies were held in the hands of foreign partners. The carbon fiber composite core aluminum conductor of Far Eastern Composite Technology Co., Ltd. adopts equivalent technical parameters to those of US CTC, and wire supporting fittings (including strain clamps, connecting pipe) will be supplied by Far East Composite Technology Co., Ltd. The wedge joint connecting fittings are imported directly from the United States.

Prior to the project establishment of the State Grid Corporation of China, the application of carbon fiber composite core aluminum conductor in the domestic power grid was completely dependent on foreign technology and high prices. The earliest application of carbon fiber composite core aluminum conductor in Fujian Province Electric Power Company reached RenMinBi (RMB) 150,000 km^{-1}, which is seven to eight times the price of regular wires. This situation seriously hindered the popularization and application of carbon fiber composite core aluminum conductor.

1.3.3 Needs of Breaking Through Construction and Operation and Maintenance Bottlenecks

The connection fittings of overseas carbon fiber composite core aluminum conductors adopt a wedge joint type structure, which is expensive. Moreover, because of excessive length and restriction of diameter, this kind of connecting hardware is difficult to pass the construction block smoothly and easily leads to breaking of the carbon fiber reinforced composite core. Therefore, the use of carbon fiber composite core aluminum conductor with wedge-type connecting fittings will adopt less 5 km section of the tension stringing method, namely the construction of long-distance lines requires frequent field transitions (including tension fields and traction fields), which greatly increases the difficulty of construction and engineering cost. The impossibility of long-distance continuous tension stringing has become one of the bottlenecks hindering the wide application of carbon fiber composite core aluminum conductors. In addition, easy installation tension full recovery preformed armor rods type connecting fittings have been widely used in the operation of conventional wire. Line operation and inspection units have proposed research and development of similar products for carbon fiber composite core aluminum conductor, to reduce the difficulty of repair operations for carbon fiber composite core aluminum conductor and workload.

1.3.4 Requirements of Promoting Development of High-Performance Carbon Fiber Industry in China

High-performance carbon fiber is an essential raw material for the aviation, aerospace, and military industries. Western countries have imposed blockades and embargos against China on the high-performance carbon fiber materials in these fields for a long time. China's research level lags far behind the advanced countries in the world, and the high-performance carbon fibers used in aircraft, missile, rocket, aerospace, weapons, and other models of structural materials are basically dependent on import. In recent years, the development of high-performance carbon fiber in China has made great breakthroughs, and many enterprises have already industrialized production of high-performance carbon fiber. This situation shows the improvement of China's carbon fiber technology level, but also indicates China's high-performance carbon fiber industry crisis. Although localization of high-performance carbon fiber is in urgent need in China's military, aerospace, and other high-tech fields, the actual amount is limited. It is difficult to maintain the existence and development of the high-performance carbon fiber industry with these limited amounts. Large and stable applications of high-performance carbon fibers must be sought in both civil and general industrial fields. The success of independent research and development of carbon fiber composite core aluminum conductor and its wide use in the power grid meet the above requirements. Therefore, taking the carbon fiber composite core aluminum conductor project as an opportunity and establishing a stable and large-scale high-performance carbon fiber industry chain has important strategic significance to promote the development of the national high-performance carbon fiber industry.

1.4 SUMMARY

Among many new types of conductors, carbon fiber composite core aluminum conductor has the advantages of lightweight, low coefficient of linear expansion (transfer temperature above), high strength, high elastic modulus, high temperature resistance, fatigue resistance, aging resistance, chemical corrosion resistance, and low resistivity (combined with soft aluminum strands) and other technical advantages. It is a new type of wire that has development potential, not only for use in improving transmission capacity and reducing power loss, but also in generally improving the line margin of safe operation. It can play an important role in the construction and upgrading of the power grid. In view of the above technical advantages of carbon fiber composite core aluminum conductor, the national Power Grid Corp in 2006 included the technology and products in its first directory of key applications of new technology.

Technology and price of the carbon fiber composite core aluminum conductor, as a new type of high-tech wire product, have been monopolized by developed countries, and the imperfection of construction and maintenance techniques constitutes two major challenges for promoting the product widely. In the above context, the "Carbon fiber composite core aluminum conductor of independent research and development and application" science and technology project was established by the State Grid Corporation of China in 2008. This project was included in the national high-tech development plan in 2010 to completely break the bottleneck of carbon fiber composite core aluminum conductor, and promote the development of the national high-performance carbon fiber industry.

CHAPTER 2

Manufacture of Carbon Fiber Composite Core Aluminum Conductor

Contents

The production of carbon fiber composite core aluminum conductor mainly includes three steps: (1) preparation of carbon fiber precursor and carbon fiber yarn; (2) production of carbon fiber composite core; (3) production of carbon fiber aluminum conductor composite core. The carbon fiber composite core is involved in the field of organic materials, which is essentially different than the steel core of conventional conductor from the material to the production thereof.

2.1 MANUFACTURE OF CARBON FIBER YARN

The core of carbon fiber aluminum conductor composite core (ACCC) is carbon fiber yarn with high strength, that is, polyacrylonitrile-based carbon fiber with a carbon content of more than 90% and high strength. The higher the grade of carbon fiber yarn, the smaller the core diameter of conductor with the same mechanical strength specification, the lighter the quality, the better the winding performance and the construction performance of the conductor. At present, carbon fiber composite core aluminum conductor self-developed domestically has begun to use high-performance carbon fiber yarn that is self-developed and produced.

Engineering Energy Aluminum Conductor Composite Core (ACCC) and its Application
DOI: https://doi.org/10.1016/B978-0-12-815611-7.00002-8

With the progress of domestic carbon fiber technology and a substantial decline in price, it does not rule out the possibility of using higher performance carbon fiber materials in the future.

The domestic carbon fiber development process is relatively tortuous. Since the 1960s, after decades of research efforts, domestic research institutes have realized in recent years that the traditional technical route for preparing carbon fiber precursor by one-step nitric acid is wrong. Shandong University Carbon Fiber Research Center has achieved a successful prototype of T700-type carbon fiber yarn in the laboratory. In 2009, relying on the project of carbon fiber composite core aluminum conductor self-developed domestically, Hebei Silicon Valley Chemical Co., Ltd. built the first domestic high-performance dry-jet wet-spinning carbon fiber precursor production line and carbonization production line. 200 t of GQ 4522-type carbon fiber yarn could be produced annually in

Figure 2.1 ***T700 carbon fiber precursor production process and product (I).*** (A) Production line panorama (tail end). (B) Production line panorama (head end). (C) Defoaming of raw materials. (D) Elongation. T700 carbon fiber precursor production process and product (II). (E) Rinse. (F) Secondary drawing. (G) Densification (hot shaft drying). (H) Heat setting. (I) Taking-up. (J) Ancillary equipment. (K) Carbon fiber yarn.

Figure 2.1 Continued

this earlier stage. (For detailed parameters, see GB/T 26752-2011 *Polyacrylonitrile-Based Carbon Fiber*, the actual antitensile strength being greater than 4600 MPa, elastic modulus being greater than 230 GPa.) Later (from 2011 to 2012), the product quality was achieved such that the variation coefficient for parameters of T700-type carbon fiber yarn was less than 5%.

Production of carbon fiber yarn includes two steps: (1) carbon fiber precursor is prepared; (2) carbon fiber precursor undergoes preoxidation and carbonization process to form carbon fiber yarn. A series of parameters and influencing factors in the production of carbon fiber precursor and carbon fiber yarn will determine the final properties of carbon fiber.

Fig. 2.1 shows a dry-jet wet-spinning carbon fiber precursor production line that completely adopts domestic technology. The production line includes the following steps: dissolution and agitation of raw materials (raw material for precursor production is added to a specific solution, dissolved, and agitated to be homogeneously mixed slurry); defoaming (defoam the slurry to remove air bubbles in the slurry to prevent breakage of precursor during spinning); spinning (spinning equipment ejects slurry to form thin yarn); drawing (the precursor is drawn so that it becomes longer and thinner and the length of the precursor is increased to four times the original one after the drawing process); rinse; secondary drawing; densification (using a heating roller to remove the moisture in the precursor to densify the precursor); heat setting (making the processed precursor retraction and setting); drying (using the heating roller to heat and dry the precursor); and taking-up (collecting carbon fiber precursor after processing), wherein the precursor spinning nozzle and other devices are self-developed domestically.

High–molecular weight polyacrylonitrile resin is prepared by continuous aqueous phase precipitation polymerization process in the production line to provide the high–molecular weight polymer needed for high-performance carbon fiber production. A high-viscosity fluid spinning assembly is designed to achieve high-speed spinning and continuous and stable operation of the dry-jet wet spinning process.

Carbon fiber precursor carbonization production includes precursor delivery, preoxidation, high-temperature carbonization, carbon fiber yarn taking-up, etc. The carbon fiber precursor carbonization production line is shown in Fig. 2.2.

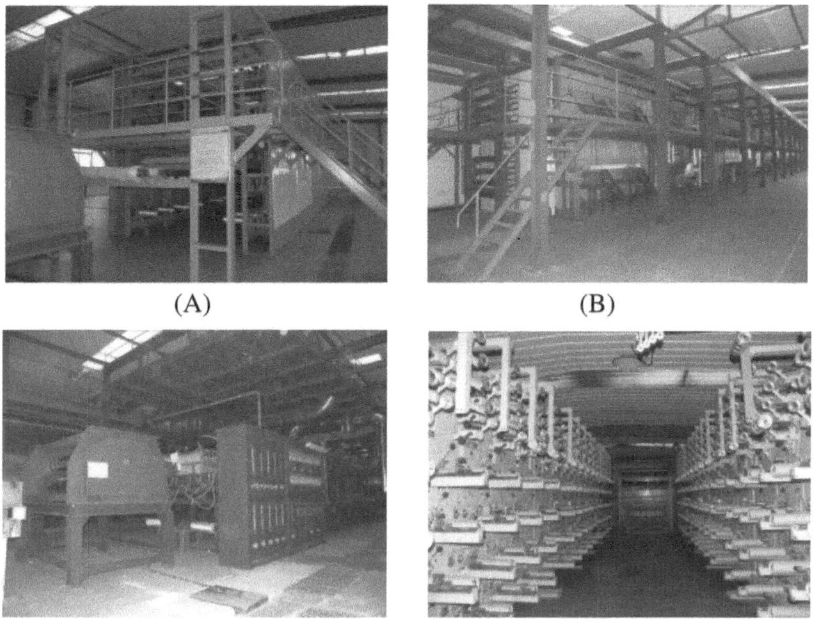

(A) (B) (C) (D)

Figure 2.2 ***Carbon fiber precursor carbonization production line.*** (A) Delivery machine. (B) Preoxidation furnace. (C) High-temperature furnace. (D) Taking-up machine.

2.2 MANUFACTURE OF CARBON FIBER COMPOSITE CORE

The carbon fiber composite core is the main load-bearing component in the carbon fiber composite core aluminum conductor, especially in high-temperature operating conditions to bear the entire conductor tension. Carbon fiber composite core reinforces resin with carbon fiber yarn, and is molded by using pultrusion process, as shown in Fig. 2.3.

Although carbon fiber yarn has the advantages of low coefficient of linear expansion (close to zero), high strength, light weight, high temperature resistance, fatigue resistance, and chemical corrosion resistance and weather aging resistance, many independent yarns are like loose sand making it difficult to for them to play a big role. Only by choosing an appropriate resin system and bonding the independent carbon fiber yarns together well to form a integrate body, that is, the carbon fiber composite core, can we maximize the synergy of carbon fiber yarns and thus give full play to the excellent performance of carbon fiber; that is, the resin system is another core of carbon fiber composite core aluminum conductors.

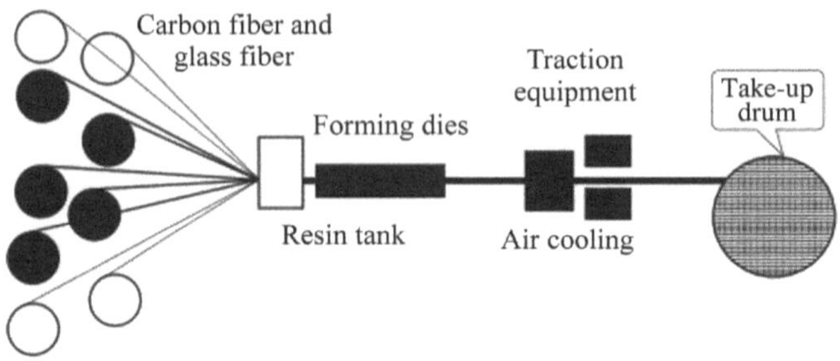

Figure 2.3 Flowchart diagram of pultrusion process.

Epoxy resin, phenolic resin, and other resin systems were tested and modified as needed, which ultimately determines the resin matrix and formulations, and has formed modified bismaleimide resin with high temperature resistance, flexibility, aging resistance, corrosion resistance, etc., the long-term heat-resistant temperature of which is up to 180°C and short-term heat-resistant temperature of which is above 200°C.

Fig. 2.4 shows the carbon fiber composite core pultrusion process production line. The production line includes dipping, plastic control, molding curing and traction, secondary curing and cutting, taking-up, and other steps.

Firstly, the carbon fiber yarn is led out of the creel and bundled into the resin tank, and the liquid resin is impregnated with the carbon fiber yarn in the resin tank to form a mixture of carbon fiber yarn and liquid resin. The resin-impregnated carbon fiber yarn penetrates the resin tank and enters into the mold. The mold inlet is tapered. Excess resin at the entrance is extruded from the carbon fiber yarns. The mold is heated and the heat is transferred to the carbon fiber yarn and the liquid resin to undergo a cross-linking reaction. The liquid resin is gradually cured from the edge of the mixture of the carbon fiber yarn and the resin to the center thereof to form a composite material of carbon fiber yarn and resin. Initially, a mixture of carbon fiber yarn and liquid resin is heated only by the temperature rise of the mold. In the latter stage, the heat released by the cross-linking reaction further heats the cured composite material so that the temperature of the composite material proceeding to the tail end of the mold exceeds the mold temperature, and therefore, appropriate measures are required to cool the composite material. The cross-linking reaction causes the volume shrinkage of the resin, separating the composite material from the mold, and the solid composite material penetrates out of

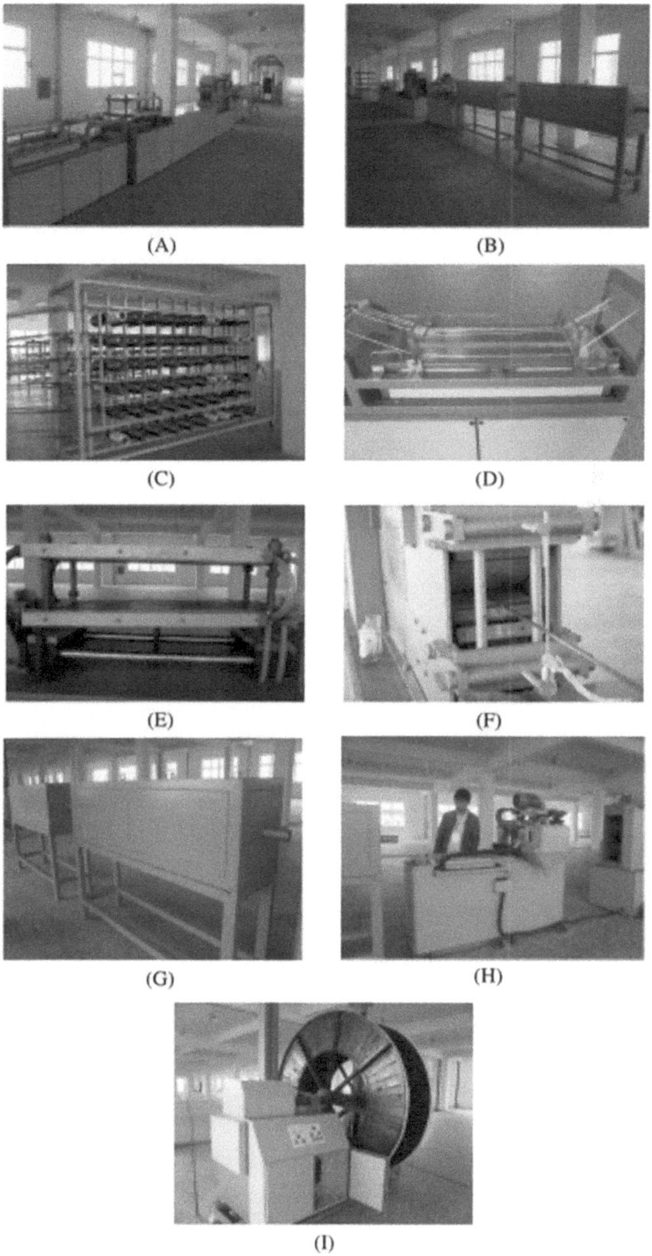

Figure 2.4 ***Carbon fiber composite core pultrusion process production line.*** (A) Production line panorama (tail end). (B) Production line panorama (head end). (C) Creel. (D) Dipping, plastic control. (E) Forming. (F) Dragging. (G) Secondary curing. (H) Cutting. (I) Wrapping up.

the mold under the action of the tractor. Finally, the composite material is cut into the finished carbon fiber composite core with the required length, and is rewound by the taking-up machine.

In the pultrusion process, the following factors can affect the final quality of the product:

1. Reinforcing material (i.e., carbon fiber yarn): carbon fiber's size, specification, and physical properties (density, lateral thermal conductivity, longitudinal thermal conductivity, specific heat, etc.), the volume content, etc.
2. Resin matrix: The chemical properties of the modified bismaleimide resin, the relationship between viscosity and temperature, the relationship between viscosity and time (such as gel time), kinetic parameters (including the total reaction heat, the reaction activation energy, the reaction level, the preconstant, etc.), the volume content of the resin system, etc.
3. Mold: Mold geometry size, mold heating zone temperature, cavity surface material and surface quality, thermal properties, etc.
4. Pulling speed: Traction speed determines the curing time of the composite material in the mold.
5. Initial temperature: The temperature of the composite material at the entrance of the mold cavity.

The carbon fiber reinforced composite material core molding process self-developed domestically introduced the non–steady state temperature model analysis method into carbon fiber composite material core preparation technology, and real-time adjustment of composite material core tensile strength can be achieved in the range of 2200–2800 MPa.

The carbon fiber reinforced composite core forming equipment self-developed domestically can realize the functions of fiber twist, fiber resin content control, fiber tension control, and fiber surface pretreatment, and increase the lateral winding of the outer layer fiber in the production line to effectively improve the longitudinal cracking resistance of the composite core.

After carbon fiber composite core self-developed domestically undergoes tens of millions of tension–tension stress fatigue tests, the residual strength can still reach 90% of the initial strength, with excellent properties of mechanical fatigue resistance. Winding performance reaches 45*D* (*D* is the diameter of composite material core), which greatly reduces the risk of splitting and wire breakage of carbon fiber composite core. The tensile strength of carbon fiber composite core at 150°C is only 5.2% lower than that at room temperature, which is much better than that of US CTC.

2.3 MANUFACTURE OF CARBON FIBER ALUMINUM CONDUCTOR COMPOSITE CORE

Aluminum strands selected on carbon fiber composite core are twisted in a certain way to form a carbon fiber composite core twisted wire. This has successfully developed and produced domestically carbon fiber composite core soft aluminum conductor to reduce conductor operating loss. It has developed and produced carbon fiber composite core heat-resistant aluminum alloy round conductor and profile line, to reduce the construction and operation overhaul difficulty of carbon fiber composite core aluminum conductors. It has developed and produced carbon fiber composite core duralumin round wire and profile lines to take into account the transmission loss and strength of aluminum strands.

2.3.1 Comparison and Selection of Aluminum Strands

Aluminum strand material, shape, and physical and chemical properties affect greatly the overall performance of the conductor (including construction, operational performance). Conventional aluminum conductor steel reinforced (ACSR) uses duralumin strand, and other new conductors (including steel heat-resistant aluminum alloy conductor, invar core conductor, aluminum-based ceramic core conductor) and commonly uses aluminum alloy strands. A comparison of aluminum strand conductive performance is shown in Table 2.1.

US CTC's carbon fiber composite core aluminum conductor uses soft aluminum strand, the main reason for which is that high-strength carbon fiber composite core can withstand all the tension of the conductor, and material with good conductivity and lower mechanical properties can be chosen with composite core conductor aluminum strand. Conductivity of soft aluminum strands of carbon fiber composite core aluminum conductors of US CTC claims to be 63.7% IACS, while that of ordinary hard

Table 2.1 Comparison of aluminum strand conductive performance

Aluminum strand materials	Monofilament conductivity	400 mm^2 nominal cross-section conductor DC resistance (Ω/km)	
	(IACS) (%)	20°C	70°C
Aluminum alloy	59–60	0.075 17	0.090 20
Duralumin	61–62	0.073 89	0.088 67
Soft aluminum	62–63	0.070 98	0.085 18

aluminum used for ACSR conductors is 61%–62% IACS and conductivity of aluminum alloy is 59%–60% IACS, that is, the soft aluminum conductivity is relatively higher than duralumin and aluminum alloy, and soft aluminum transmission loss is relatively lower than duralumin and aluminum alloy. Therefore, currently domestic experts tend toward the application of soft aluminum strand conductor. However, since the mechanical properties (hardness, tensile strength, etc.) of the soft aluminum strands are much lower than that of the conventional aluminum strands, the tensile strength of the soft aluminum strands is 60 M–90 MPa (only 50% of the hard aluminum strength). During construction and operation, it is easy to deform and damage the conductor (see Fig. 2.5), which makes the construction and operation more difficult. For example, due to the deforming of the aluminum strands during the delivery, the accessory must be added to effectively control the relative displacement between the composite core and the aluminum strands. The conductor passes through a number of trolleys; especially when the corner is too large it is prone to strand mismatch deformation. During the process of tighten the conductor, the commonly used and easy to operate gripper has to be given up and a cumbersome pretwisted conductor gripper is used. The use of outlet aerodyne during operation and maintenance can easily cause aluminum strand deformation and so on.

Conventional conductors use round aluminum strands, and soft aluminum strands are specially shaped aluminum, such as trapezoidal soft aluminum strand carbon fiber composite core aluminum conductors in foreign technology and Z-shaped (or S-shaped) soft aluminum strand carbon fiber composite core aluminum conductors in domestic technology

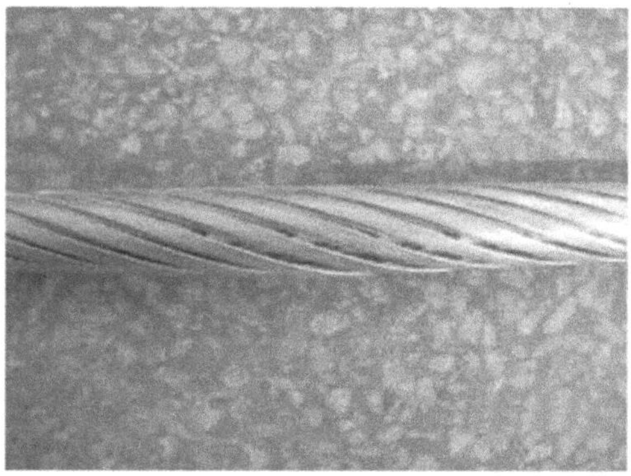

Figure 2.5 Carbon fiber composite core soft aluminum profile conductor with deformation of aluminum strand.

(see Fig. 2.6). This form eliminates the gaps between the aluminum strands and objectively increases the cross-sectional area of the aluminum strand current carrying conductors under the same conductor outer diameter, which is conducive to operation under high winds and icing conditions. One of the original motivations for using special-shaped aluminum may be to fill up the gap between the soft aluminum strands to avoid more deformation of the soft aluminum strands when squeezed.

The safety and reliability of the conductors and transmission losses are both important factors to be considered when choosing aluminum strands. At present, many new conductors (thermal-resistant aluminum alloy conductor (TACSR), zirconium alloy aluminum conductor invar steel reinforced, aluminum conductor composite reinforced (ACCR)) use aluminum alloy strands with high mechanical strength and slightly lower conductivity, while US carbon fiber ACCC uses soft aluminum strands with significantly lower mechanical strength and slightly higher conductivity, each focusing on different items, but both obviously not the best choice. Generally it is to be selected according to the actual operating conditions; for example, when the need for lower line loss during operating and better operating environment and the probability of occurrence of conductor damage is low, the preferred choice is of carbon fiber composite core soft aluminum conductor. When construction conditions are difficult and operating environment is poor, such as with lines undergoing frequent conductor damage, the preferred choice is of carbon fiber composite core heat-resistant aluminum alloy conductor. If the reliability and loss of conductors cannot be taken into account at the same time, you should give priority to safety and reliability; in Japan, where land is scarce,

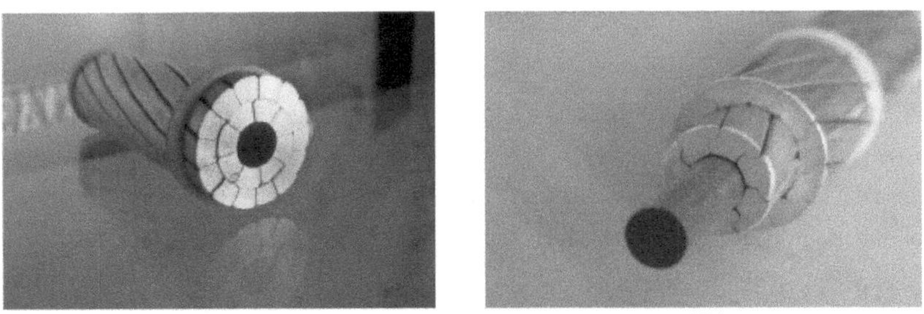

(A) (B)

Figure 2.6 ***Carbon fiber composite core aluminum conductor with trapezoidal soft aluminum strands and Z-shaped soft aluminum strands.*** (A) Trapezoidal soft aluminum strand. (B) Z-shaped soft aluminum strand.

the preferred choice is high-strength heat-resistant aluminum alloy strand for trial of carbon fiber composite core aluminum conductor, instead of using the more conductive soft aluminum strand. In fact, the use of soft aluminum strand instead of the ordinary hard aluminum strand in common conductors may lead to loss decreased by about 5%, which is not very impressive. Carbon fiber composite core aluminum conductor is known as a "high temperature augmented capacity" conductor; once operating at 150°C or higher, the loss due to temperature increase is far more than the high conductivity of soft aluminum can offset, so that loss reduction function of carbon fiber composite core aluminum conductor only makes sense when the soft aluminum strands must be used and conductor operating temperature thereof is similar to that of the conventional conductor.

For the carbon fiber composite core aluminum conductor with higher strength aluminum strand, heat-resistant aluminum alloy strands or hard aluminum strands were selected respectively according to the differences in operating conditions.

1. For the conductors whose maximum operating temperature is greater than 90°C, the heat-resistant aluminum alloy is selected for the aluminum strands of the carbon fiber composite core aluminum conductor to ensure the higher mechanical strength of the aluminum strands at normal temperature and high temperature and the stability of the conductor parameters after high-temperature operation. The application of carbon fiber composite core heat-resistant aluminum alloy conductor in 10–500 kV overhead transmission lines shows that this type of conductor has great advantages in construction and operation overhaul. Compared with soft aluminum strand, the aluminum alloy strand can effectively prevent the deformation and damage of the soft aluminum strands during construction and operation, and reduce the construction difficulty.

The use of heat-resistant aluminum alloy strands leads to an increase in the temperature at the migration point of the carbon fiber composite core aluminum conductor. Table 2.2 is for the JLNR/F2B-300/65 carbon fiber composite core aluminum alloy conductor technical parameters table, the conductor migration point temperature is in the range of 100–110°C, higher than the carbon fiber composite core soft aluminum conductor migration point temperature (about 80°C). However, in applications of a number of augmented capacity lines, the stress–sag calculation results show that the carbon fiber composite core heat-resistant aluminum alloy conductor can meet the sag requirements after augmented capacity

Table 2.2 JLNR/F2B-300/65 carbon fiber composite core aluminum alloy conductor technical parameters

Item		Unit	Product warranty value	
Structure	Aluminum profile conductor	Piece	26	
	Composite core	Piece	1	
Nominal cross-section	Aluminum profile conductor	mm^2	299.5	
	Composite core	mm^2	62.0	
	Comprehensive cross-section	mm^2	361.5	
Twisting direction of the outer layer			Right	
Outer diameter		mm	24.17	
Calculated mass		kg/km	920	
Ultimate tensile strength		kN	180	
Elastic modulus		GPa	72.4 (below 110°C)	178 (above 110°C)
Linear expansion coefficient		10^{-6}/°C	12.8 (below 110°C)	2.0 (above 110°C)
20°C DC resistance (calculated value)		Ω/km	0.0989	
Aluminum conductor 20°C resistivity		Ω mm^2/m	0.028 750 5	
Allowable temperature for long-term operating		°C	150	
Allowable maximum operating temperature (short time)		°C	180	

because the sag advantage of the carbon fiber composite core aluminum conductor comes from the two aspects of light weight and high-temperature linear expansion coefficient, and the effect of light weight is more prominent in most occasions.

In the selection and test of aluminum strands, it is found that the temperature at the migration point (inflection point temperature) of the carbon fiber composite core aluminum conductor is not a constant. As the aluminum strand hardness and mechanical strength increase, the temperature at the migration point of the conductor also increases (for example, the difference between aluminum alloy strands and soft aluminum strands). As the ratio of the aluminum strand cross-section to the composite core cross-section increases, the temperature at the migration point of the conductor increases. The temperature at the migration point of the domestically produced soft aluminum round conductor strand carbon fiber composite core aluminum conductor is even as low as 60°C (see Table 2.3), which is 20°C lower than that of the carbon fiber composite core soft aluminum conductor of US CTC. However, after comprehensive consideration, the conductor is not put into practical use.

Conventional hard aluminum strand used in ACSR belongs to nonannealing aluminum, while soft aluminum strand belongs to completely annealed aluminum. Hard aluminum after high-temperature operation (such as 150°C) is equivalent to a certain degree of annealing; mechanical strength and hardness are decreased, and the conductivity is increased. According to regulations, the mechanical strength reduction of aluminum strands in the service life (according to the 30-year depreciation period) is not allowed to exceed 10%, so if the duralumin is in regular high-temperature operation (such as 150°C), the degree of mechanical strength decline will not meet the above requirements. However, if the long-term operating temperature of duralumin is not more than 90°C, the mechanical strength of aluminum strands and other parameters will not be significantly reduced. Taking into account the transmission loss should not be too large, in the line design, maximum operating temperature of the conductor is generally not more than 70°C, and the maximum operating temperature of some lines having augmented capacity requirements will not exceed 90°C. Therefore, it is proposed that the conductors of maximum operating temperature should be not higher than 90°C; hard aluminum strands used in the conventional ACSR are selected as aluminum strands of carbon fiber composite core aluminum conductors to form carbon fiber composite core hard aluminum strands to achieve the

Table 2.3 Carbon fiber composite core soft aluminum conductor parameters for comparative tests (ACCC/RL-240/45 model)

Item		Unit	Value	
Soft aluminum single conductor	The number/diameter of strands at the first layer	Piece/mm	9/3.70	
	The number/diameter of strands at the second layer	Piece/mm	16/3.41	
Carbon fiber core conductor	Pieces/diameter	Piece/mm	1/7.5	
Nominal cross-section	Aluminum conductor	mm^2	242.8	
	Composite core	mm^2	44.2	
	Comprehensive cross-section	mm^2	287.0	
Conductor outer diameter		mm	21.72	
Soft aluminum single conductor conductivity			≥ 63% IACS	
Calculated mass		kg/km	757.2	
Ultimate tensile strength		kN	120	
Elastic modulus		GPa	69.6 (below 60°C)	178 (above 60°C)
Linear expansion coefficient		10^{-6} °C	12.7 (below 60°C)	1.9 (above 60°C)
20°C DC resistance (calculated value)		Ω/km	0.1131	
Aluminum conductor 20°C resistivity		Ω mm^2/m	0.02731	

Note: ACCC/RL-240/45 conductor is only carbon fiber composite core soft aluminum conductor used for comparative tests. Because the soft aluminum strands used are round conductor strands, they are not used in the soft aluminum conductor specification of Q/GDW 1851-2012 *Carbon Fiber Composite core Overhead Conductor*. Therefore, the temporary model ACCC/RL is used in the text.

2. For the circuit with conductor maximum operating temperature not more than 90°C, the selection of hard aluminum strands as carbon fiber composite core aluminum conductor aluminum strands is for the formation of carbon fiber composite core hard conductor to take into account the higher mechanical strength of aluminum strands and higher conductivity and low operating loss.

balance between the mechanical strength of aluminum strands and the transmission loss of the conductors, and to ensure that the aluminum mechanical strength does not decline after a long-term operation below 90°C and the conductor elasticity modulus, linear expansion coefficient, and other conductor parameters remained stable.

[*Case*] Taking the 500-kV Shachang double-circuit line augmented capacity project in North China as an example, the augmented capacity conductor is selected and relevant line parameters are calculated.

1. Selection of conductor. Without grasping the maximum operating temperature of augmented capacity conductors, it is temporarily recommended to use carbon fiber composite core heat-resistant aluminum alloy twisted conductors in the augmented capacity and reconstruction project of the Shachang line. To match the original line use of LGJ-300/40 ACSR (parameters in Table 2.4), it is recommended to use JLNR/F2B-300/65 carbon fiber composite core heat-resistant aluminum alloy twisted conductors; for detailed parameters see Table 2.2.
2. Calculation of transmission capacity and operating temperature. Taking four splitting JLNR/F2B-300/65 carbon fiber composite core heat-resistant aluminum alloy conductors as an example, calculate the conductor transmission capacity:

Table 2.4 LGJ-300/40 ACSR parameters

Comprehensive cross-sectional area (mm^2)	338.99	Unit quality (kg/km)	1133
Calculation of outer diameter (mm)	23.94	Calculated tensile force (kN)	92.2
Aluminum strand number/diameter per strand (mm)	24/3.99	Elastic modulus (kN/mm^2)	73
Steel strand number/diameter per strand (mm)	7/2.66	Temperature linear expansion coefficient of 1°C	19.6×10^{-6}
Steel strand total gross section (mm^2)	300.09	Maximum imposed tension (kN)	35
Steel core total cross-section (mm^2)	38.90	Safety factor	2.5

Largest load of 500 kV Shachang line No. 1 and No. 2 is 1,500,000 kW; the maximum load of 2,500,000 kW under $N-1$; corresponding maximum phase current of 2887 A (four splitting conductor).

According to the ambient temperature of 40°C, wind speed of 0.5 m/s and sunlight intensity of 1000 W/m^2, the maximum operating temperature of JLNR/F2B-300/65 carbon fiber composite core heat-resistant aluminum alloy conductor is 90°C. In accordance with the maximum operating temperature, the carbon fiber composite core aluminum twisted conductors can also be applied to the augmented capacity of the line.

3. Calculation of stress–sag. Taking JLNR/F2B-300/65 carbon fiber composite core heat-resistant aluminum alloy conductor as an example, calculate the conductor stress–sag. The average operating tension and the maximum operating tension of the conductor are the same as those of the original ACSR. The meteorological conditions are calculated according to the Beijing section of the original Shachang line, as shown in Table 2.5.

The calculation results show that under the conditions of the conductor tension not exceeding the original ACSR, under various operating conditions (including the conductor maximum operating temperature, in this case 90°C) carbon fiber composite core aluminum conductor sag is no more than that of the original ACSR line (see Table 2.6).

Table 2.5 Meteorological conditions calculation

Meteorological condition		Temperature (°C)	Wind speed (m/s)	Thickness of ice (mm)
Maximum temperature		40	0	0
Minimum temperature		−30	0	0
Annual average temperature		5	0	0
Maximum wind speed		−5	32	0
Icing		−5	10	5, 10
Atmospheric overvoltage	Breezeless	15	0	0
	Windy	15	10	0
Operation overvoltage		5	16	0
Installation		−15	10	0
The number of thunderstorm days		40 days/year		

Table 2.6 Stress–sag calculation results

LGJ-300/40 model	Representative span	Minimum temperature		Maximum wind speed		Maximum temperature		Icing		Conductor 70°C	
		Stress (MPa)	Sag (m)	Stress (MPa)	Sag (m)	Stress (MPa)	Sag (m)	Stress (MPa)	Sag (m)	Stress (MPa)	Sag (m)
	100	103.35	0.40	78.60	0.85	32.00	1.28	81.66	0.94	22.38	1.83
	200	97.21	1.68	90.29	2.96	45.71	3.58	96.59	3.17	36.78	4.45
	300	84.27	4.37	94.60	6.36	51.88	7.10	103.35	6.67	44.98	8.19
	400	71.17	9.20	93.04	11.49	53.10	12.34	103.35	11.86	48.35	13.55
	500	64.99	15.75	92.14	18.13	53.78	19.03	103.35	18.53	50.37	20.32
	600	61.81	23.85	91.60	26.27	54.19	27.20	103.35	26.68	51.64	28.54
JLNR/F2B-300/65 type	Representative span	Minimum temperature		Maximum wind speed		Maximum temperature		Icing		Conductor 90°C	
		Stress (MPa)	Sag (m)	Stress (MPa)	Sag (m)	Stress (MPa)	Sag (m)	Stress (MPa)	Sag (m)	Stress (MPa)	Sag (m)
	100	89.89	0.35	76.00	0.78	36.70	0.85	77.84	0.84	20.66	1.52
	200	82.94	1.51	85.50	2.76	44.78	2.80	89.50	2.90	32.47	3.86
	300	72.86	3.87	90.85	5.84	48.45	5.82	96.49	6.06	39.38	7.15
	400	60.95	8.22	89.73	10.52	47.69	10.50	96.49	10.78	41.77	11.99
	500	55.30	14.15	89.06	16.55	47.29	16.55	96.49	16.84	43.17	18.13
	600	52.39	21.51	88.65	23.95	47.05	23.95	96.49	24.25	44.05	25.58

In summary, both the carbon fiber composite core soft aluminum twisted conductor and the carbon fiber composite core heat-resistant aluminum alloy twisted conductor can operate at a temperature of 150°C or more, but the aluminum strands of the soft aluminum twisted conductor have low mechanical strength and are not conducive to construction and operation. Aluminum strands of heat-resistant aluminum alloy twisted conductor have slightly lower conductivity and slightly more loss. Some lines of the actual line augmented capacity renovation project only need the conductor operating temperature increased from 70°C to 90°C and below to meet the augmented capacity requirements, so the use of carbon fiber composite core duralumin twisted conductor can fully meet the operational requirements, and can take into account the aluminum strand mechanical strength and transmission loss.

2.3.2 Twisting of Aluminum Strands

2.3.2.1 Carbon Fiber Composite Core Heat-Resistant Aluminum Alloy Twisted Conductor

Carbon fiber composite core heat-resistant aluminum alloy twisted conductor production equipment and process can be divided into three parts.

1. Electrolysis furnace and aluminum rod continuous casting and milk equipment: After aluminum ingot and the intermediate alloy are melted in the electrolytic furnace, they are then subjected to heat preservation alloying, and then continuously casted and milked to form an aluminum alloy rod after refining and stirring.
2. Drawer: The aluminum alloy rod undergoes drawing annealing to form the desired size aluminum alloy strands.
3. Conductor frame twisting machine: Twisting aluminum alloy strands on the composite core and forming a carbon fiber composite core heat-resistant aluminum alloy twisted conductor.

Fig. 2.7 is a production line for carbon fiber composite core heat-resistant aluminum alloy twisted conductor by twisting heat-resistant aluminum alloy round conductor strands. The production line is similar to that of the conventional hard aluminum ACSR.

2.3.2.2 Carbon Fiber Composite Core Soft Aluminum Twisted Conductor

China has developed a composite core trapezoidal and Z-shaped (or S-shaped) soft aluminum single-conductor drawing technology, softening technology, and stranded technology.

Figure 2.7 ***Stranding of carbon fiber composite core heat-resistant aluminum alloy twisted conductor.*** (A) Production line panorama. (B) Composite core shaft. (C) Installing the composite core. (D) Installing aluminum strand shaft. (E) Stranding the first layer of aluminum strands. (F) Stranding the second layer of aluminum strands. (G) Guide wheels. (H) Wrapping up.

Conventional vertical conductor drawing machines cannot guarantee the drawing of trapezoidal soft aluminum monofilament, and are prone to stand up and scratch phenomena. China uses the self-designed sliding horizontal conductor drawing machine, and the drawing process needs a circulating cooling device. High quality lubricant gives good lubrication to stretching metal and molding the hole surface; in the meantime to form a film layer can withstand the pressure in the mold hole without destruction, which can reduce the friction coefficient, ensure profile lines with surface smooth, shape regular and accurate size, and match good cooling cycle device. Using frequency control technology controls the ratio, the tension controller makes the profile line tension uniform, the polycrystalline drawing die ensures that the structure of the profile conductor monofilament stays within rules and has a smooth surface, and the automatic nondestructive taking-up device makes the aluminum strand drawing process continuously stable, to ensure that all aspects of the profile line are not damaged.

Softened aluminum is made of high-purity electrical aluminum from heating, boring fire, exhaust, tempering, and other processes at a specific time and temperature. To ensure uniform softening of profile conductors, the design of the skeleton plate in the softening process, the use of advanced temperature control system, uniform temperature field, and ventilation effect, are controlled so that the line is evenly softened inside and outside, and softened aluminum strength is kept within the required range. After softening conductivity is increased to 63%, IACS strength is 80 MPa, to ensure that the conductor has good ductility and electrical conductivity.

During the stranding process, profile conductors are prone to turning over, and the line is not round, the surface scratches, and other issues. At trial production, to ensure the length of the product, the design of more aluminum single conductors, there is a single conductor uneven force problem in the flange.

In response to this problem, the redesign of the flange of the stranding equipment, according to the product structure to design, equally divided flange, to solve the problem. Other problems to solve include profile conductors turning over and riding and unrounding of the conductor; the domestic independent design and manufacture of the sublocating mold, locator, conductor loop, arched conductor core support, double flapper mold, and forming regulator; oil purifying device turning over, and turning and other technical problems in the conductor stranding process (Fig. 2.8).

Figure 2.8 Stranding of carbon fiber composite core soft aluminum twisted conductor.

2.3.3 Serialization of Carbon Fiber Composite Core Aluminum Conductor

Conventional conductors are serialized based on the cross-sectional area of aluminum strands and steel cross-sectional area, for example, LGJ-400/35-type conductor denotes ACSR with steel core cross-sectional area of 35 mm^2 and aluminum strand cross-sectional area of 400 mm^2, the conductor series division of which is currently widely accepted; see GB/T 1179-2008 *Round Conductor Concentric Stranding Overhead Conductor.* However, the series division of carbon fiber composite core aluminum conductor cannot be completely copied ACSR, which is because the steel core is twisted by multiple round steel strands, and the current mainstream of the carbon fiber composite core is in single form, that is, the steel core and the composite core of same cross-sectional area have different outer diameters, or the steel core and the composite core having the same outer diameter have different cross-sectional areas. US CTC's carbon fiber composite core aluminum conductor series is divided according to the British cross-sectional area, which does not meet China's national conditions. Therefore, in the initial domestic research and development of carbon fiber composite core aluminum conductor, there was no unified carbon fiber composite core aluminum conductor series, but a more consistent view is that the resulting series of products are neither suitable for being completely in accordance with the conventional conductor nor suitable for being completely in accordance with foreign technology of carbon fiber composite core aluminum conductor division.

Finally, according to the outer diameter of carbon fiber composite core to divide composite core series, including the series of 5.0, 5.5, 6, 6.5, 7, 7.5, 8, 8.5, 9, 9.5, 10, 10.5, 11.0 mm (13 series in total), the current series division has been included in GB 29324-2012 *Overhead Conductor Fiber Reinforced Resin-based Composite Mandrel* and Q/GDW 1851-2012 *Carbon Fiber Composite Core Overhead Conductor*. If a higher mechanical strength carbon fiber composite core is required, it is not recommended to use a single-core method, otherwise the winding and construction performance of the composite core will be reduced. Other methods may be considered, such as multistrand fine composite core stranding into higher strength composite core, similar to the structure of conventional steel core. Taking JLNR/F2B-300/65 carbon fiber composite core heat-resistant aluminum alloy twisted conductor as an example, composite core outer diameter is of 9 mm, and aluminum strand cross-sectional area is of 300 mm^2. The conductor composite core diameter of 9 mm is corresponding to the actual composite core cross-sectional area of 63 mm^2, but corresponds to the same conductor outer diameter of the conventional steel core cross-sectional area of only 50 mm^2. Therefore, some experts have suggested marking based on direct composite core diameter, such as JLNR/F2B-300/9.0, but, after repeated trade-offs, settled on finally marking as an integer closest to the actual cross-sectional area of the composite core in Q/GDW 1851-2012, such as JLNR/F2B-300/65.

2.4 SUMMARY

1. The first dry-jet wet-spinning high-performance carbon fiber precursor industrialized production line and its supporting carbonization production line have been developed independently in China, and the production lines can produce annually 200 t of T700 carbon fiber yarn. Also developed independently have been a carbon fiber composite core pultrusion—winding process, equipment, and the modified bismaleimide resin with high temperature stability, and carbon fiber composite core has been manufactured successfully, key indicators of which (high-temperature mechanical strength and antifracturing performance) are significantly better than similar foreign products, and production base of carbon fiber composite core for conductor with an annual capacity of 20,000 km has been formed.
2. The trapezoidal soft aluminum single-wire drawing technology and stranded process have been developed successfully, and also soft

aluminum profile wire series of carbon fiber composite core aluminum conductor has been developed and applied. Heat-resistant aluminum alloy round wire strand series of carbon fiber composite core aluminum conductor has been developed and applied, which effectively reduces the difficulty of construction and operation and maintenance of the carbon fiber composite core aluminum conductor. Heat-resistant aluminum alloy profile wire strand carbon fiber composite core aluminum conductor wire has been trial-manufactured to improve operating performance of the conductor under wind and icing conditions. Hard aluminum strand carbon fiber composite core aluminum conductor has been trial-manufactured to take into account the mechanical strength and operating loss of aluminum strand.

3. Selection principle for aluminum strand of carbon fiber composite core aluminum conductor is proposed: The soft aluminum twisted wire is preferred when low wire loss and better operating environment are required during operation and the probability of conductor damage is lower. Heat-resistant aluminum alloy twisted wire is preferred when construction conditions are difficult, the operating environment is poor, the conductor is easily damaged, and the maximum operating temperature of the conductor is greater than 90°C. Hard aluminum twisted conductor can be chosen when the conductor is easily damaged and when the maximum operating temperature of the wire is not greater than 90°C, to take into account the mechanical strength and operating loss of aluminum strands.

CHAPTER 3

Carbon Fiber Composite Core Aluminum Conductor Detection

Contents

Carbon fiber composite core aluminum conductor detection items, methods, and equipment are different from that of conventional conductors. The detection thereof includes two parts. (1) Conventional parameter test. Carbon fiber composite core aluminum conductor should meet the requirements for mechanical and electrical performance of conventional conductor, that is, it should pass through a full set of conventional conductor type test items. (2) Artificial accelerated aging test and mechanical fatigue test. The carbon fiber composite core is an organic material, which is an innovation for both conductor manufacturing enterprises and application units. Its antiaging performance has drawn wide attention. State Grid Hebei Electric Power Research Institute (Hebei Electric Power Research Institute) has drawn lessons from organic composite insulators artificial accelerated aging test widely used in the grid and combined with the characteristics of the conductor to design and implement carbon fiber composite core aluminum conductor composite factor and an artificial accelerated aging test, thus laying a foundation for the safe operation and standard setting of carbon fiber composite core aluminum conductor.

Engineering Energy Aluminum Conductor Composite Core (ACCC) and its Application
DOI: https://doi.org/10.1016/B978-0-12-815611-7.00003-X

In addition, the threat of alternating/impact mechanical load on the material is much higher than the static mechanical load, which easily leads to mechanical fatigue damage and destruction of the material. Overhead lines operating environment varies widely, and mechanical fatigue damage due to breeze vibration, dancing, and other alternating mechanical stress is a problem that must be faced in conductor operation. Performance of carbon fiber composite core withstanding long-term mechanical fatigue is one of the issues that most users focus on. Northern China Electric Power Research Institute simulated various forms of mechanical vibration that may be encountered in the operation of carbon fiber composite core aluminum conductor, and implemented for the first time a carbon fiber composite core aluminum conductor's 12×10^7 breeze vibration fatigue test, 3×10^7 splitting conductor vertical vibration fatigue test, horizontal vibration fatigue test, and torsional vibration fatigue test, giving a more comprehensive verification that the carbon fiber composite core aluminum conductor has mechanical fatigue resistance performance that is not less than or even better than ordinary aluminum conductor steel reinforced (ACSR), that is, considering from mechanical fatigue resistance, carbon fiber composite core aluminum conductor can operate safely for the long term on overhead transmission lines. In addition, some units have carried out research on the detection methods of composite core defects during the manufacture or construction of carbon fiber composite core aluminum conductors. For example, State Grid Smart Grid Research Institute has attempted detection of small composite core defects at the exit of the tensile machine using X-ray. In addition, the State Grid Liaoning Electric Power Supply Co., Ltd. has attempted to increase the microdefects in the composite core manufacturing process before the taking-up process, combined with the inspection of the composite core of inner and outer layers insulating voltage, hoping that defect-free carbon fiber composite core is produced and used for overhead transmission lines.

3.1 CONVENTIONAL PARAMETER TEST

The carbon fiber composite core aluminum conductor self-developed domestically has passed the conventional conductor parameter tests of Shanghai Electric Cable Research Institute and China Electric Power Research Institute (formerly Beijing Electric Power Construction Research Institute), including carbon fiber composite core heat-resistant aluminum alloy twisted conductor JLNR/F2B-300/65 type (9 mm

Table 3.1 JLNR/F2B-300/65 carbon fiber composite core heat-resistant aluminum conductor size parameters

Aluminum alloy single conductor	The number/diameter of strands at the first layer	Piece/mm	10/3.83
	The number/diameter of strands at the second layer	Piece/mm	16/3.83
Carbon fiber core conductor	Pieces/diameter	Piece/mm	1/8.95
Conductor outer diameter		mm	24.17
Pitch diameter ratio	Inner layer aluminum strand	13.8	
	Outer layer aluminum strand	10.7	

Table 3.2 JLNR/F2B-240/45 carbon fiber composite core heat-resistant aluminum conductor size parameters

Aluminum alloy single conductor	The number/diameter of strands at the first layer	Piece/mm	9/3.70
	The number/diameter of strands at the second layer	Piece/mm	16/3.41
Carbon fiber core conductor	Pieces/diameter	Piece/mm	1/7.5
Conductor outer diameter		mm	21.72

composite core, see Table 3.1 for dimension parameters), JLNR/F2B-240/45 type (7.5 mm composite core, see Table 3.2 for dimension parameters), etc. The test reference standard is GB/T 1179-1999 *Round Conductor Concentric Strand Overhead Conductor*, of which JLNR/F2B-300/65 carbon fiber composite core heat-resistant aluminum alloy twisted conductor has been firstly successfully applied to ultrahigh voltage (UHV) overhead lines—500 kV Wanshun line No. 3. Table 3.3 shows the dimensional parameters of ACCC/RL-240/45 soft carbon round conductor strand carbon fiber composite core aluminum conductor (7.5 mm composite core) for comparison.

1. Conductor grip test at room temperature (see Fig. 3.1)

   The tensile force of JLNR/F2B-240/45 carbon fiber composite core heat-resistant aluminum alloy conductor at room temperature is

Table 3.3 ACCC/RL-240/45 soft aluminum round conductor strand carbon fiber composite core aluminum conductor size parameters

Soft aluminum single conductor (Round aluminum strand)	The number/diameter of strands at the first layer	Piece/mm	9/3.70
	The number/diameter of strands at the second layer	Piece/mm	16/3.41
Carbon fiber core conductor	Pieces/diameter	Piece/mm	1/7.5
Conductor outer diameter		mm	21.72

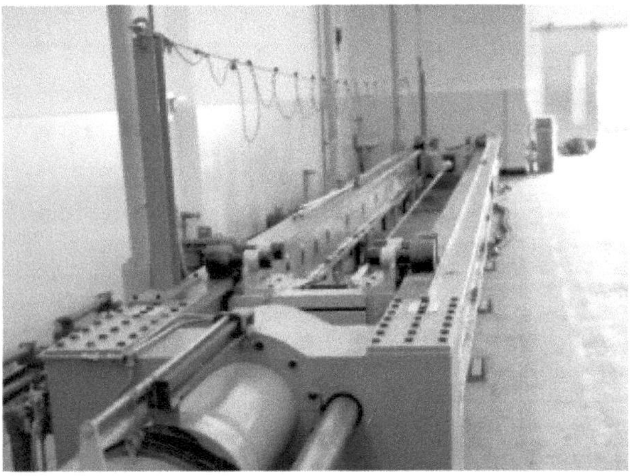

Figure 3.1 Test of conductor tensile force and strain clamp grip strength.

165.6 kN, and the tensile force of ACCC/RL-240/45 round conductor strand carbon fiber composite core soft aluminum conductor at room temperature is 159.7 kN, while the calculated tensile force of the two kinds of conductors is 124.29 kN, the measured result of which is higher than the calculated tensile force by more than 30%.

2. Breaking load test of conductor at high temperature (see Fig. 3.2)

   Tensile force of JLNR/F2B-240/45 heat-resistant aluminum alloy conductor at room temperature is 165.6 kN, tensile force thereof at 150°C is 149 kN, and the tensile force thereof at 170°C is 149.6 kN. Its ratio of high temperature tensile force thereof at 150°C and 170°C to tensile force at room temperature is 90%, that is, it has excellent high temperature resistance performance.

Figure 3.2 Test of tensile force at high temperature.

3. Carbon fiber composite core tensile force test

The 7.5-mm composite core of JLNR/F2B-240/45 type heat-resistant aluminum alloy conductor and ACCC/LR-240/45 type soft aluminum conductor has a tensile force of 123.7 kN at room temperature, and tensile strength of the composite core at room temperature is 2800 MPa, which is similar to foreign products.

9.0-mm composite core used in JLNR/F2B-300/65-type heat-resistant aluminum alloy conductor has a tensile force of 167 kN at 150°C, and tensile strength thereof is 2656 MPa. Therefore, when the temperature rises from room temperature (at 30°C) to 150°C, the tensile strength of self-developed carbon fiber composite core decreased by only 5.2%.

Tensile strength of US CTC's carbon fiber composite core at 160°C is only 1400 MPa. When the temperature is increased from room temperature (at 30°C) to 150°C, the tensile strength of the composite core is reduced by more than 40%.

One of the important advantages of carbon fiber composite core aluminum conductor is to improve line transmission capacity. The key to improve line transmission capacity is to raise the operating temperature of the conductor, that is, to maintain the conductor's higher mechanical strength at high temperature and to make conductor parameters unchanged after operation at high temperature, so tensile force at high temperature is one of the most important indicators of carbon fiber composite core aluminum conductor. Test data shows that in terms of indicator of tensile force at high temperature, the carbon fiber composite core aluminum conductor self-developed domestically is superior to foreign products.

4. Conductor mass

Conductor mass per unit length is the main parameter that determines the initial sag of the conductor. The mass per unit length of JLNR/F2B-300/65 carbon fiber composite core heat-resistant aluminum alloy twisted conductor is 920.35 kg/km and the rated tensile force is 180 kN. The mass per unit length of LGJ-300/50 ACSR of the same size is 1210 kg/km, and rated tensile force is 103.4 kN. The mass of JLNR/F2B-300/65 carbon fiber composite core aluminum conductor is only 76% of conventional conductor LGJ-300/50 type, and if you consider the advantages of carbon fiber composite core aluminum conductor mechanical strength, you can consider using a conductor with smaller composite core cross-section, thus the mass advantage determines the significant advantage of the carbon fiber composite core aluminum conductor in stress–sag. Taking into account the vast majority of conductor operating time being within room temperature range, to a certain extent the mass advantage for reducing the role of carbon fiber composite core aluminum conductor sag exceeds the role of low linear expansion coefficient at high temperature.

5. Conductor stress–strain properties

The stress–strain characteristic of the conductor is one of the important parameters for designing the sag of the conductor, which reflects the elastic elongation of the conductor when it is under mechanical load such as its own weight and ice coating. The test chart thereof is shown in Fig. 3.3. The JLNR/F2B-300/65 conductor has a final elasticity modulus of 72.4 GPa, which is similar to conventional ACSR.

Figure 3.3 Conductor stress–strain characteristics test.

Figure 3.4 Composite core crush bending resistance test.

Figure 3.5 Composite core crush resistance performance test.

6. Composite core antidamage performance

Composite core bending resistance performance test (see Fig. 3.4) and crush resistance performance test (see Fig. 3.5) are test items proposed for carbon fiber composite core. The above two properties of composite core developed domestically are significantly better than those of similar foreign products. Take JLNR/F2B-300/65 conductor as an example:

1. The maximum crush resistance load of the composite core developed domestically is 34.7 kN, while the maximum crush resistance load of a composite core in a US company is about 25 kN.
2. The maximum bending load of composite core developed domestically is 4.7 kN, while the maximum bending load of composite core in a US company is about 2.5–3.5 kN.

Compared with the steel core of conventional conductors, carbon fiber composite core is a brittle material. When the applied pressure is too large, the composite core may have fragmentation, so crimping connection fittings are used for carbon fiber composite core aluminum conductor instead of wedge connection fittings, and the composite core developed domestically has a significant advantage.

7. Composite core winding test

Composite core winding test (see Fig. 3.6) is a test item proposed for carbon fiber composite core; the smaller the diameter of the winding circle, the less likely it is to break, and the better the workability. Fig. 3.7 shows that the bending phenomenon in the construction of carbon fiber composite core aluminum conductor, Q/GDW 1851-2012 "carbon fiber composite material core overhead conductor" requires that the diameter of the winding circle is not more than $55D$ ($D =$ composite core diameter). The diameter of the winding circle for JLNR/F2B-300/65 style composite core (diameter of 9 mm) reaches 50 mm, and the diameter of the winding circle for JLNR/F2B-240/45 type composite core (diameter of 7.5 mm) reaches 40 mm, so as to meet the standards and on-site construction need. For the composite core whose diameter is 9 mm, the tensile force thereof has reached 150 kN, which can meet the application requirements of general UHV and extrahigh voltage lines, it is proposed that composite core with greater tensile force no longer uses a single core type, so as to avoid winding performance decreasing with the increasing of composite core diameter, which is not conducive to construction and operation.

Figure 3.6 Composite core winding test.

Figure 3.7 Bending phenomenon of carbon fiber composite core conductor during construction.

Figure 3.8 Conductor sag–temperature characteristics test.

8. Conductor sag–temperature characteristic test and linear expansion coefficient test

Conductor sag–temperature characteristics test (see Fig. 3.8) and linear expansion coefficient test (see Fig. 3.9) are two of the main parameters involved in conductor design, reflecting the characteristics of the conductor sag changing with temperature. Carbon fiber composite core aluminum conductor has migration point temperature (temperature inflection point) in the allowable operating temperature range, below the migration point temperature. Carbon fiber composite core aluminum conductor linear expansion coefficient is similar to the conventional conductor, wherein linear expansion coefficient of JLNR/F2B-300/65 carbon fiber

Figure 3.9 Linear expansion coefficient test.

composite core heat-resistant aluminum alloy twisted conductor is 12.8×10^{-6} (1/°C), and linear expansion coefficient of ACCC/RL-240/45 carbon fiber composite core soft aluminum twisted conductor is 12.7×10^{-6} (1/°C). Above the migration point, the linear expansion coefficient of the carbon fiber composite core aluminum conductor decreases drastically. The linear expansion coefficient of JLNR/F2B-300/65 is 2.0×10^{-6} (1/°C) and the linear expansion coefficient of ACCC/RL-240/45 is 1.9×10^{-6} (1/°C), so that the sag of the conductor basically does not increase during operating at high temperature. Therefore, the linear expansion coefficient above the temperature of the migration point is one of the key indicators of the carbon fiber composite core aluminum conductor, which ensures the advantage of low sag of the conductor together with the low mass of the carbon fiber composite core aluminum conductor.

Carbon fiber yarn linear expansion coefficient is close to zero, but after forming the composite core by combining with the resin and then forming the carbon fiber composite core aluminum conductor by combining with the aluminum strands, the overall linear expansion coefficient of the conductor is determined by the carbon fiber yarn, resin, and aluminum strand. With the change of the properties of the resin and aluminum strands and the change of the cross-section ratio of aluminum strands and composite core, the temperature of the migration point of the conductor also shows a certain difference.

The use of aluminum alloy strands with higher hardness and mechanical strength will raise the migration point temperature of carbon fiber composite core aluminum conductor. For example, the migration point temperature of JLNR/F2B-300/65 carbon fiber composite core heat-resistant aluminum alloy twisted conductor is greater than 100°C, the migration point temperature of US CTC's carbon fiber soft aluminum twisted conductor is 80°C, and the migration point temperature of ACCC/RL-240/45 carbon fiber composite core soft aluminum twisted conductor trial-manufactured domestically for comparison test is 60°C.

9. Conductor DC resistance test at 20°C

DC resistance is related to conductor loss, ampacity, and other performance indicators. A conductor DC resistance test is shown in Fig. 3.10. DC resistance at 20°C of JLNR/F2B-300/65 carbon fiber composite core heat-resistant aluminum alloy twisted conductor is 0.0989 Ω/km, which is higher than carbon fiber composite core soft aluminum twisted conductor of the same size by about 5%; that is, its operating loss is slightly higher than the soft aluminum twisted conductor, which is determined by the difference between the conductivity of heat-resistant aluminum alloy and soft aluminum.

10. Resistivity test of aluminum single conductor

The aluminum single conductor resistivity is a test item for aluminum strands (see Fig. 3.11), which is also a parameter related to the performance of conductor loss and ampacity. Resistivity at 20°C of JLNR/F2B-300/65 heat-resistant aluminum alloy twisted conductor is 0.0287505 Ω mm^2/m, which is about 5% higher than the soft aluminum twisted conductor. The test result of aluminum single conductor resistivity is consistent with the conductor DC resistance test result.

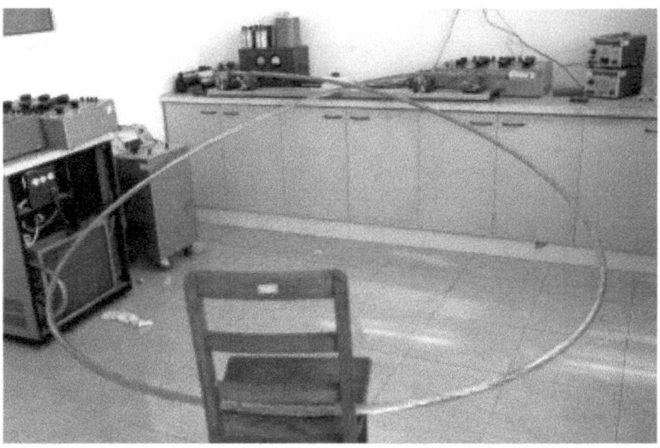

Figure 3.10 Conductor DC resistance test.

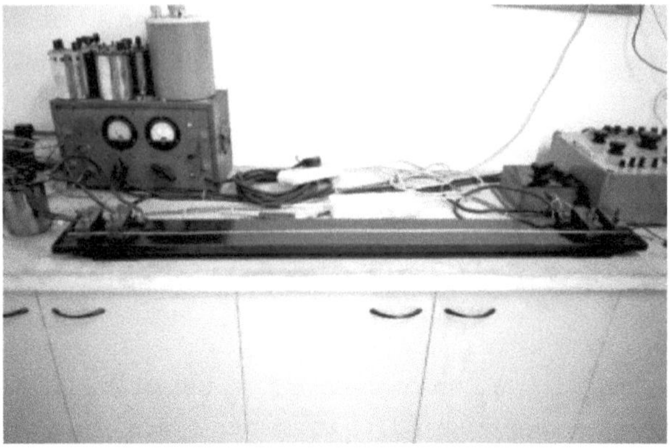

Figure 3.11 Aluminum single conductor resistivity test.

Figure 3.12 Aluminum single conductor tensile force test.

11. Aluminum single conductor tensile force test

Aluminum single conductor tensile force is a test item for aluminum strand (see Fig. 3.12). Hard aluminum strands used in conventional ACSR have a tensile strength of 175.9 MPa. The tensile strength of the soft aluminum strands used in the ACCC/RL-240/45 type is 78 MPa (60 M–95 MPa), which is more than 50% lower than that of the duralumin, i.e., the damaged and deformation probability of soft aluminum

Figure 3.13 Creep test at room temperature.

conductor during construction and operation may be higher than that of the conventional conductor.

The tensile strength of JLNR/F2B-300/65 heat-resistant aluminum alloy strands is 166 MPa, which is similar to the strength of hard aluminum strands. Therefore, JLNR/F2B-300/65 type heat-resistant aluminum alloy strands are much better than soft aluminum strands in terms of avoiding damages and deformations during construction and operation and reducing difficulties of conductor construction and maintenance.

12. Room/high temperature creep test

The creep test of the conductor at room temperature (see Fig. 3.13) reflects the amount of plastic deformation of the conductor during long-term operation, and the parameter relates to the change of sag of the conductor. Creep tests of carbon fiber composite core aluminum conductor include two sets of tests at room temperature and high temperature (150°C), and each set of tests respectively tests the creep amount of the test pieces under 15%, 25%, and 35% of the rated tensile force. The creep amount of 10 years of JLNR/F2B-300/65 carbon fiber composite core heat-resistant aluminum alloy twisted conductor at room temperature under 25% of the rated tensile force is 328 mm/km, which is 34% lower than that of US CTC's carbon fiber composite core aluminum conductor and also lower than the conventional ACSR. For the plastic deformation of the conductor, it is generally processed by cooling method in the circuit design. In the application of carbon fiber composite core aluminum conductor, when processing conductor plastic elongation by cooling

Figure 3.14 Fatigue vibration test with self-developed connecting fittings.

method, it should take into account of this difference in creep; for example, cooling by 20°C is replaced by cooling by 10°C or 15°C.

13. Fatigue vibration test

Metal materials are often able to withstand greater static stress, but are easily damaged under the action of smaller alternating stress, such as strand breakage of the conductor under the breeze vibration, or even conductor breakage. Fatigue vibration test simulates the breeze vibration that may be experienced during conductor operating (see Fig. 3.14) to examine the ability of the conductor to withstand this alternating stress.

Multiple models of carbon fiber composite core heat-resistant aluminum alloy twisted conductor developed domestically have passed the 30 million fatigue vibration test of Shanghai Cable Research Institute, and there is no aluminum strand breakage and composite core damage.

14. Test of passing the tackle

Tension release is an important part of the construction of the line. One end of the cable is connected to the tension machine during releasing (see Fig. 3.15), one end of the cable is connected to the tractor, and the middle of the cable is passed through the pay-off tackle on each tower (see Fig. 3.16). The conductor should be released with a certain tension, to maintain a certain distance with the crossing substance.

When the conductor is released with tension, the conductor will be wound when the conductor passes through the tensioner axle and the tackle, and if the parameters are not properly selected, the conductor may

Figure 3.15 Tension machine for construction of releasing with tension.

Figure 3.16 Carbon fiber composite core conductor passing through the double pulley.

be damaged due to a plurality of factors such as conductor tension, envelope angle, and tackle diameter during the process of passing the tackle. The composite core of carbon fiber composite core aluminum conductor belongs to organic composite material composed of fiber and resin; its axial tensile force is much better than steel core (steel twisted conductor), but the flexibility is lower than normal steel core, which also ensures that passing through the tackle when releasing the conductor with the tension is different from the conventional conductor and needs to conduct a comprehensive test (test equipment is shown in Fig. 3.17).

Figure 3.17 Test equipment for passing the tackle.

When the conductor passes the tackle, the factors that may cause its damage mainly include:

1. Tackle diameter. The greater the tackle diameter, the less likely it is that the conductor will be damaged when it passes the tackle.
2. Conductor tension. The greater the conductor tension is, the greater the possibility of damage to the aluminum strands and carbon fiber composite core will be. Although a winding test is required for the composite core, such as that a composite core of 9 mm is capable of winding a circle of diameter of 50 mm, the test is performed under tension-free, static conditions on the conductor, and the conductor is subjected to tension of tens of tons and in the moving state when conducting the release with tension. Therefore, the tackle and tensioner axles need a larger diameter to ensure that the carbon fiber composite core is not damaged when the conductor is released with tension.

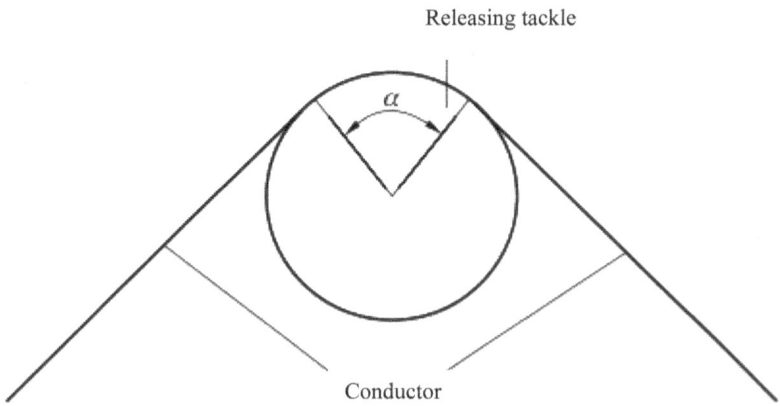

Figure 3.18 Envelope angle of the conductor passing the pulley.

3. Envelope angle. The envelope angle (angle α as shown in Fig. 3.18) corresponds to the central angle of the arc of the tackle enveloped by the conductor. The greater the envelope angle is, the greater the possibility of damage to the aluminum strands and carbon fiber composite core will be.

4. The number of times passing the tackle. The greater the number of times passing the tackle, the greater the possibility of damage to the aluminum strands and the carbon fiber composite core will be.

5. Other factors. Releasing speed, the material of the tackle (mainly the inner surface state of the tackle groove), etc. all have an effect on when the conductor passes the tackle.

To conduct a comprehensive test of passing the tackle for the carbon fiber composite core aluminum conductor, the conditions of the carbon fiber composite core aluminum conductor passing the releasing tackle under various influencing factors are investigated to determine the control parameters of passing the releasing tackle, and a test system for the conductor passing the releasing tackle is designed. The test system may simulate the conditions of field releasing with tension of carbon fiber composite core aluminum conductor. By loading the tension and making the conductor pass the releasing tackle repeatedly, the aluminum strand and core material status before and after the conductor passing the tackle are compared.

The test system for the conductor passing the releasing tackle mainly consists of four parts: (1) infrastructure, (2) releasing tackle, (3) hydraulic loading system, and (4) human–machine interface (i.e., software control

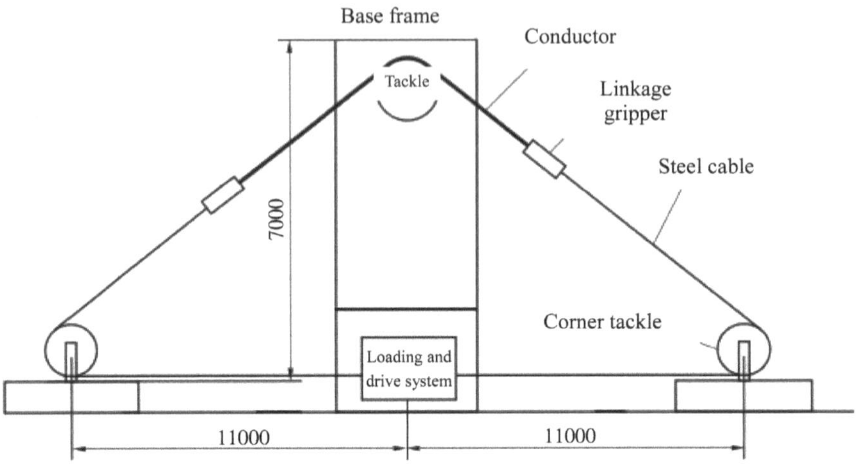

Figure 3.19 Structure diagram of test system for conductor passing the releasing tackle.

system). Test instructions are issued by the human–machine interface, and transmitted by the program to the electrohydraulic servo system for conductor loading. When the conductor is added to a given load, the system begins to automatically test, and it stops automatically after the test ends, that is, the test process is fully automated after the test conductor is installed. The system structure diagram is shown in Fig. 3.19.

1. The conductor tension is adjustable and constant during the test, and the maximum tensile that can be loaded by the test system is 50 kN.
2. The envelope angle of the conductor passing the tackle is adjustable. The range of the envelope angle that the test system can simulate is 15–60 degrees.
3. The form of tackle is adjustable. The test system can be disassembled separately, and single tackle, double tackles, and tackle blocks can be specially designed and erected according to the need to simulate different construction requirements and working conditions, as shown in Fig. 3.20.

To determine a reasonable tackle size, pay-off tension, and the envelope angle when passing the tackle, a comprehensive test for passing the releasing tackle on the carbon fiber composite core aluminum conductor developed domestically is conducted using the test system for the conductor passing the releasing tackle (see Fig. 3.21). During the test, the conductor tension, envelope angle and number of times passing the tackle are

Figure 3.20 Tackle.
(A) Single tackle. (B) Double tackle. (C) Tackle block.

Figure 3.21 Test for carbon fiber composite core aluminum conductor passing the releasing tackle.
(A) Test for double tackles. (B) Test for tackle blocks.

changed, and the constricted condition of the conductor is observed at the same time.

The larger the diameter of the carbon fiber composite core, the larger the minimum bending radius allowed, the more difficult it is to pass the tackle. At present, the maximum diameter of composite core of the carbon fiber composite core aluminum conductor that is actually put into use is 9 mm. Therefore, the test is performed on a JLNR/F2B-300/65 carbon fiber composite core heat-resistant aluminum alloy twisted conductor as a sample with a diameter of 9 mm, and the test uses high-strength nylon wheel tackle.

1. Tackle 660
   a. The test is carried out by a tackle 660 with a groove bottom diameter of 660 mm. The length of the conductor is more than 10 m. The envelope angle of the conductor passing through the tackle is 32 degrees. Both ends of the conductor are tensioned with a conductor gripper and both ends of the conductor are equipped with spinner. The conductor tension is 10 kN. The conductor is reciprocated on the tackle 20 times (i.e., the conductor passes the tackle 40 times) without obvious damage to the aluminum strands. The aluminum strands are peeled off to inspect the composite core, the core bar has no breakage and no damage, and there is no significant displacement between the conductor aluminum strands and the composite core, that is, there is no obvious core shrinking phenomenon.
   b. The test is carried out by a tackle 660 with a groove bottom diameter of 660 mm. The length of the conductor is more than 10 m. The envelope angle of the conductor passing through the tackle is 32 degrees. Both ends of the conductor are tensioned with a conductor gripper and both ends of the conductor are equipped with spinner. The conductor tension is 12 kN. The conductor is reciprocated on the tackle 20 times (i.e., the conductor passes the tackle 40 times) without obvious damage to the aluminum strands. The aluminum strands are peeled off to inspect the composite core, the core bar has no breakage and no damage, and there is no significant displacement between the conductor aluminum strands and the composite core, that is, there is no obvious core shrinking phenomenon.
   c. The test is carried out by a tackle 660 with a groove bottom diameter of 660 mm. The length of the conductor is more than 10 m.

The envelope angle of the conductor passing through the tackle is 32 degrees. Both ends of the conductor are tensioned with a conductor gripper and both ends of the conductor are equipped with spinner. The conductor tension is 15 kN. When the conductor is reciprocated on the tackle 12 times (i.e., the conductor passes the tackle 24 times), the conductors are accompanied by a breaking sound, but the aluminum strands have no significant damage. The aluminum strands are peeled off to inspect the composite core, and the core bar has a fracture point.

2. Tackle 822
   a. The test is carried out by a tackle 822 with a groove bottom diameter of 822 mm. The length of the conductor is more than 10 m. The envelope angle of the conductor passing through the tackle is 45 degrees. The conductor tension is 15 kN. The conductor is reciprocated on the tackle 30 times (i.e., the conductor passes the tackle 60 times) without obvious damage to the aluminum strands. The aluminum strands are peeled off to inspect the composite core, the core bar has no breakage and no damage, and there is no significant relative displacement between the conductor aluminum strands and the composite core, that is, there is no core shrinking phenomenon.
   b. The test is carried out by a tackle 822 with a groove bottom diameter of 822 mm. The length of the conductor is more than 10 m. The envelope angle of the conductor passing through the tackle is 32 degrees. The conductor tension is 20 kN. When the conductor is reciprocated on the tackle 24 times (i.e., the conductor passes the tackle 48 times), the conductors are accompanied by a breaking sound, but the aluminum strands have no significant damage. The aluminum strands are peeled off to inspect the composite core, and the core bar has a fracture point.

To sum up, for carbon fiber composite core heat-resistant aluminum alloy twisted conductor (such as JLNR/F2B-300/65 type) with a composite core diameter of 9 mm, a tackle with a diameter of not less than 800 mm (such as a tackle with a groove bottom diameter of 822 mm) should be selected for the tension stringing, and envelope angle (approximately equal to line turning angle) of the conductor passing a single tackle should not be greater than 30 degrees; double tackles should be used when the envelope angle is greater than 30 degrees. As for construction machinery of carbon fiber composite core aluminum conductor with composite core diameter of less than 9 mm, please select by referring to the parameters above. In addition, carbon fiber composite core heat-resistant aluminum alloy twisted conductor has no obvious shrinking core, which reduces the construction difficulty of carbon fiber composite core aluminum conductor. Shrinking core phenomenon easily occurs in construction such as when releasing with tension, tightening the conductor, etc. of carbon fiber composite core soft aluminum twisted conductor, so some measures should be taken to prevent core shrinking.

3.2 ARTIFICIAL ACCELERATED AGING TEST

Artificial accelerated aging test involves applying stress (including electrical, mechanical, temperature, rainfall, solar radiation, etc.) above the normal operating condition on the sample, to explore the antiaging properties of the sample in a relatively short period of time and calculate its running life. The following issues should be noted in the design of an artificial accelerated aging test:

1. The stress applied is properly increased on the basis of normal stress, such as appropriate increase of voltage amplitude, and proper extension of irradiation intensity and duration, but excessive increase of severeness should be avoided. For example, the test temperature should be raised to near the glass temperature of conductor composite core and so on. Excessive increase of the test conditions may lead to great difference between the destruction principle of the sample and the aging and destruction principle of the sample in actual operation, and thus the reference meaning is lost.
2. The difference between single factor and composite factor: In actual operation, the product often bears a variety of stress at the same time, and damage and aging of some products are due to a variety of stresses, that is, aging and the destruction effect under a variety of stresses are greater than the effect under a single stress, such as that zero suspension disk insulators are generally caused by voltage and tension. The single operating voltage leads to a low probability of zero insulators. For example, the insulator strings in a substation structure bear less tension and the zero value rate is relatively low.

This section consists of single-factor artificial accelerated aging test and composite factor artificial accelerated aging test.

Figure 3.22 UV irradiation test equipment and samples.
(A) Testing equipment. (B) Sample.

3.2.1 Single-Factor Artificial Accelerated Aging Test

1. Ultraviolet irradiation test

An ultraviolet irradiation test simulates ultraviolet in daylight to irradiate carbon fiber composite core aluminum conductor to investigate the antiultraviolet radiation aging performance of carbon fiber composite core (see Fig. 3.22). Comparing the bending performance and crush resistance of composite core samples after being irradiated by UV light and not irradiated by UV light, the samples irradiated by UV light are divided into composite core with aluminum and composite core without aluminum to investigate the blocking effect of aluminum strand on the composite core.

a. UV irradiation process. The process of UV irradiation for 4 h and condensation for 4 h is cycled; the cycle process is performed for 15 days in total. The initial UV irradiance is 0.119 mW/cm^2. As the UV lamp ages, the final UV irradiance is decayed to 0.067 mW/cm^2.

b. Composite core bending failure test (see Fig. 3.23).

i. Composite core sample without aluminum strands: The length of the sample is 100 mm, the diameter D is 7.5 mm, and the test span is 6D (45 mm). After 15 days of UV aging, the bending damage forces of the five composite cores are 2.41, 2.43, 2.40, 2.53, and 2.30 kN, respectively, and average value thereof is 2.41 kN.

ii. Composite core sample with aluminum strands: The length of the sample is 100 mm, the diameter D is 7.5 mm, and the test span is 6D (45 mm). After UV aging for 15 days, the aluminum strands outside the composite core are peeled off, and the bending

(A) (B)

Figure 3.23 Composite core three-point bending test.
(A) Testing equipment. (B) Sample.

(A) (B)

Figure 3.24 Composite core crush failure test.

damage forces of the two composite cores are 2.34 and 2.28 kN, respectively, and average value thereof is 2.31 kN.

iii. Composite core samples without undergoing UV irradiation: The length of the sample is 100 mm, the diameter D is 7.5 mm, and the test span is $6D$ (45 mm). The bending damage forces of the two composite cores are 2.28 and 2.32 kN, respectively, and average value thereof is 2.30 kN.

c. Composite core crush failure test (see Fig. 3.24). The length of the sample is 50 mm, a diameter D is 7.5 mm. As for samples with

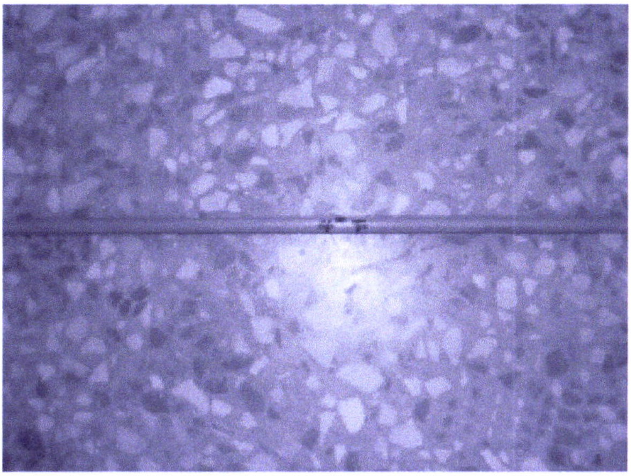

Figure 3.25 Composite core without undergoing UV radiation.

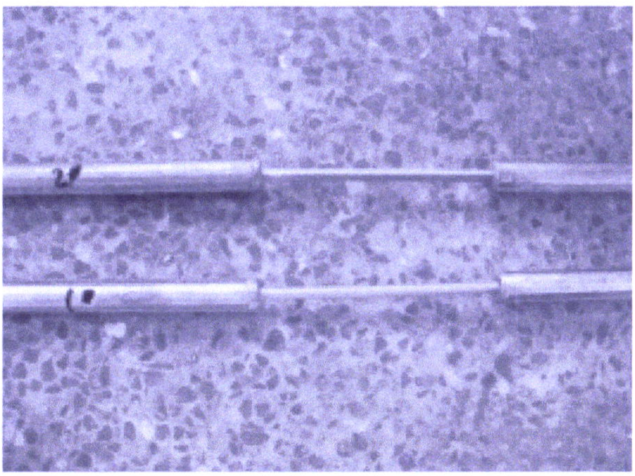

Figure 3.26 Composite core undergoing UV radiation.

undergoing UV aging for 15 days and samples not undergoing UV aging, the crush forces of the composite cores are both greater than 10 kN.

The composite cores before and after UV irradiation are shown in Figs. 3.25 and 3.26, respectively. The contrast of sample surface colors shows the following. As for samples aged for 15 days by ultraviolet light, the surface color of the composite cores becomes noticeably lighter, but aluminum strands help to shield the composite cores from UV radiation.

Figure 3.27 Low temperature test equipment.

The composite core bending failure test and crush failure test show that composite core damage resistance performance has no significant decline before and after UV aging.

2. Hot and cold cycle test

a. Hot and cold cycle test conditions. The carbon fiber composite core sample is placed in an environment of −25°C for 48 h (low temperature test equipment is shown in Fig. 3.27), and then the sample is removed and placed in a room temperature environment for 24 h. Many hot and cold cycles are conducted like this, and then bending failure and tensile failure performance of composite cores with and without hot and cold cycle are compared.

b. Composite core bending failure test.

i. Hot and cold cycles are conducted eight times: The length of the sample is 100 mm, the diameter D is 7.5 mm, and the test span is $6D$ (45 mm). After hot and cold cycling, the bending damage forces of the two composite cores are 2.79 and 2.20 kN, respectively, and an average value thereof is 2.50 kN.

ii. Without undergoing hot and cold cycle: The length of the sample is 100 mm, the diameter D is 7.5 mm, and the test span is $6D$ (45 mm). The bending damage forces of the two composite cores are 2.28 and 2.32 kN, respectively, and average value thereof is 2.30 kN.

c. Composite core tensile force test.

i. Hot and cold cycles are conducted four times: The carbon fiber composite core is crimped with a supporting steel anchor; this composite core undergoes four hot and cold cycles. The tensile force test is performed, and the tensile force of the composite core is 106 kN.

ii. Hot and cold cycles are conducted eight times: The carbon fiber composite core is crimped with a supporting steel anchor; this composite core undergoes eight hot and cold cycles. The tensile force test is performed, and the tensile force of the composite core is 102 kN.

iii. Without undergoing hot and cold cycles: The carbon fiber composite core is crimped with a supporting steel anchor, then the tensile force test is performed directly on this composite core, and the tensile force is 108 kN.

The composite core bending failure test and tensile force test show that the composite core damage resistance performance has no significant decline before and after hot and cold cycle.

3. Acid solution corrosion and salt solution corrosion test

The carbon fiber composite core is immersed in acid (sulfuric acid) and salt (sodium chloride) reagent (see Figs. 3.28 and 3.29) with a concentration of 3% respectively and stands at the laboratory standard temperature (25 ± 2°C) for more than 24 h. The composite core is removed for

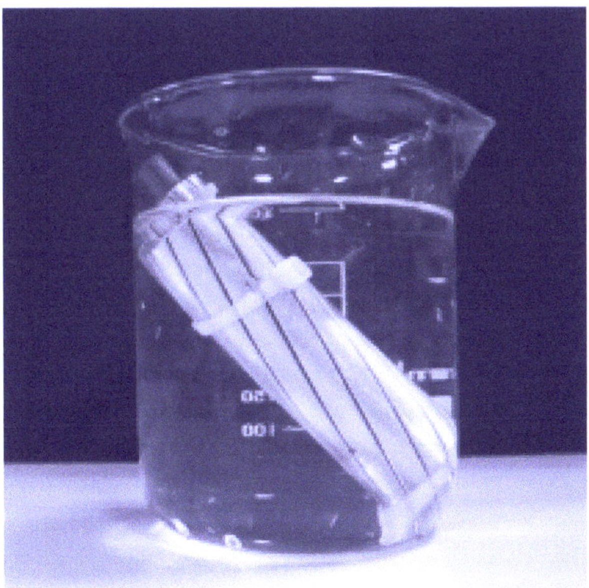

Figure 3.28 Composite core salt corrosion test.

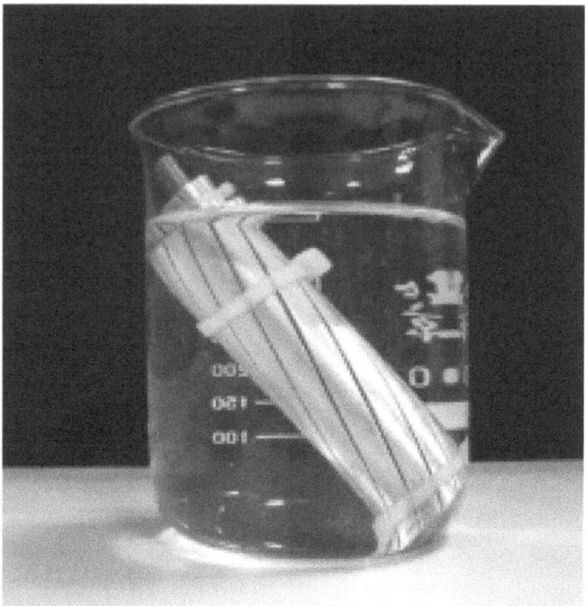

Figure 3.29 Composite core acid corrosion test.

cleaning, and composite core bending failure test and crush failure test are performed. The test indicated that the composite core damage resistance performance before and after acid solution and salt solution corrosion test has no significant decline.

3.2.2 Composite Factor Artificial Accelerated Aging Test

3.2.2.1 Composite Factor Artificial Accelerated Aging Test Method

Composite factor artificial accelerated aging test on the composite organic insulators (briefly referred to as composite factor aging test) is relatively mature. There are four kinds of composite factor aging test methods worldwide, including IEC[1] 5000 h composite factor aging test, ENEL[2] 5000 h composite factor aging test, EPRI[3] composite factor aging test (divided into summer/winter cycle aging test and desert climate aging test), and FGH[4] 5000 h composite factor aging test.

[1] International Electrotechnical Commission (IEC).
[2] Ente Nazionale per l'Energia eLettrica (ENEL).
[3] Electric Power Research Institute.
[4] German Power System and Electricity Economy Joint Research Society.

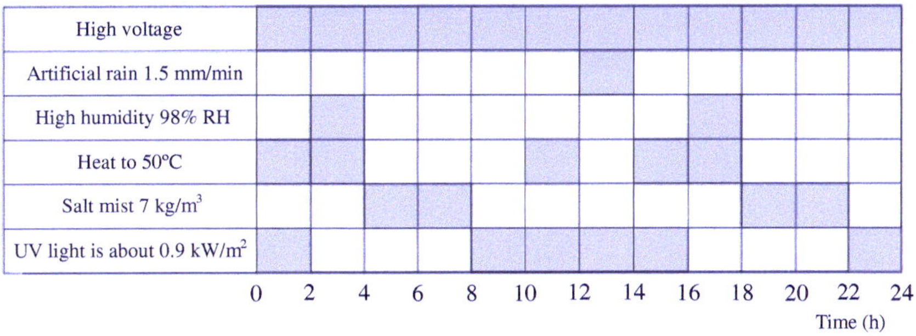

Figure 3.30 IEC composite factor aging test procedure.
Note: Shading means applying factors in this time period, the same as below.

1. IEC 5000 h composite factor aging test

The composite factor aging test method proposed in Annex 3 of IEC 62217 *Polymer High Voltage Insulators for Indoor and Outdoor* was introduced in the late 1960s. After more than 10 years of development it was finalized in the early 1980s, and the international council on large electric systems (CIGRE)[5] 5000 h composite factor aging test is similar to it. The composite factor aging procedure of IEC is shown in Fig. 3.30.

As with the artificial climate accelerated aging test, IEC 5000 h composite factor aging test focuses on simulating natural climate change with strong pertinence. As shown in Fig. 3.30, after the UV light (simulated sunlight), the sample is warmed (heated) and then is fogged up. At this time, the UV light is hard to transmit to the Earth's surface, so the UV light source is not turned on and the sunlight then reappears (i.e., the UV light source is turned on), the sample is heated, and then it rains, followed by hot and humid weather (heated and high humidity), and then salt spray and UV light are applied to begin the next cycle. As can be seen from this process, the simulation of this climate tends more to be a tropical or subtropical climate.

2. ENEL 5000 h composite factor aging test

The ENEL 5000 h composite factor aging test parameter is similar to the IEC 5000 h composite factor aging test, except that salt content in the salt spray is increased to 80 kg/m^3, UV intensity is raised from 0.9 to 1.5 kW/m^2, and the cycle period is extended to a cycle of 7 days. The test procedure is shown in Fig. 3.31.

The test cycle period extension of ENEL 5000 h composite factor aging test is beneficial to the simulation of hydrophobic loss and recovery

[5] CIGRE.

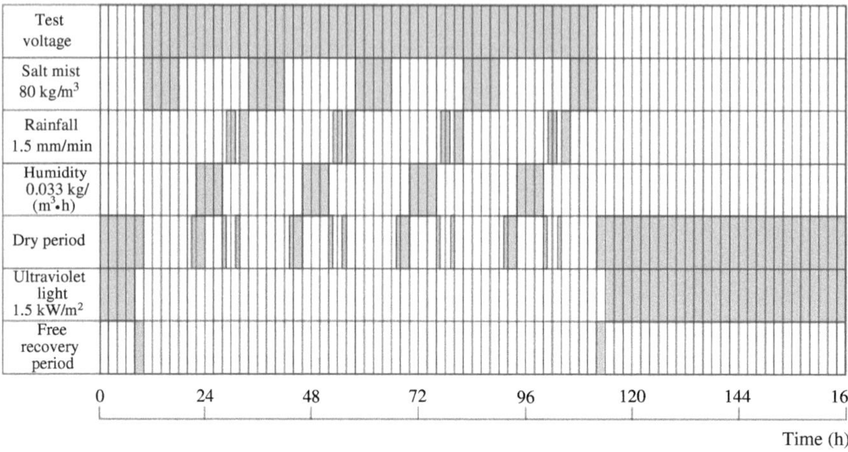

Figure 3.31 ENEL 5000 h composite factor aging test procedure.

of silicone rubber composite insulators, but the salt content of 80 kg/m^3 in the contamination test is a heavy contamination. This heavy contamination without insoluble salt results in the hydrophobicity of the sample in the bottom umbrella surface and rod stem and other areas that are difficult to wash by the rain are almost impossible to recover, which will inevitably make the sample undergo serious discharge under high concentrations of salt spray. Therefore, the test method is very different from the contamination situation of composite insulator on-site operation, and it is also very different in the aging model.

3. EPRI composite factor aging test

EPRI composite factor aging test methods are divided into two kinds: one is the summer/winter cycle aging test, and the other is an aging test for the desert climate. Summer/winter cycle aging tests are tailored to the type of climate in the coastal areas of the state of Florida in the United States. The two different procedures, cycled for 24 h, are included in the test: one is the summer procedure, and the other is winter procedure, which are shown in Figs. 3.32 and 3.33, respectively. Salt spray only appears in the winter procedure. To simulate the weak discharge in actual operation, the salt content in salt spray is dropped to 2.5 kg/m^3. The summer procedure takes into account higher temperatures and more rainfall and UV light.

The desert climate aging test is formulated for the climate in the western United States. It mainly features less rainfall, very slight contamination, and longer sunlight exposure. The test procedure is shown in Fig. 3.34.

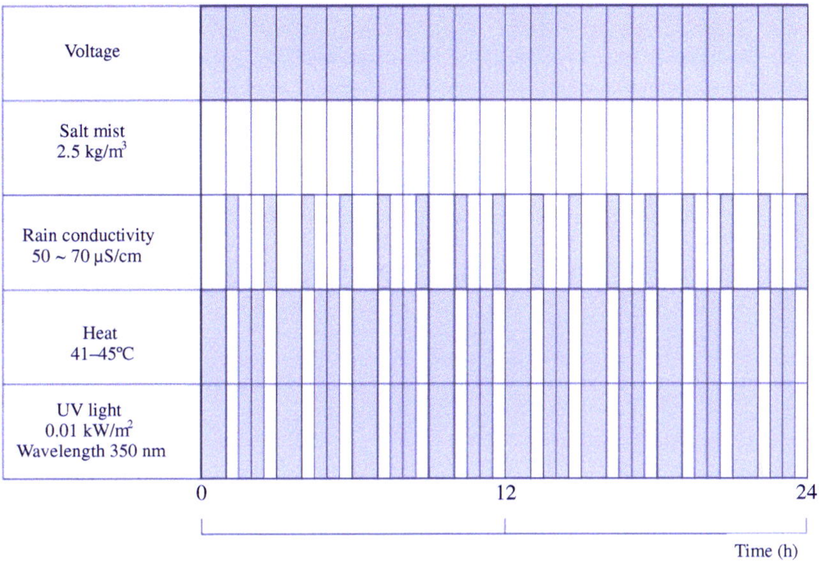

Figure 3.32 Summer cycle aging test procedure in EPRI summer/winter aging cycle procedure.

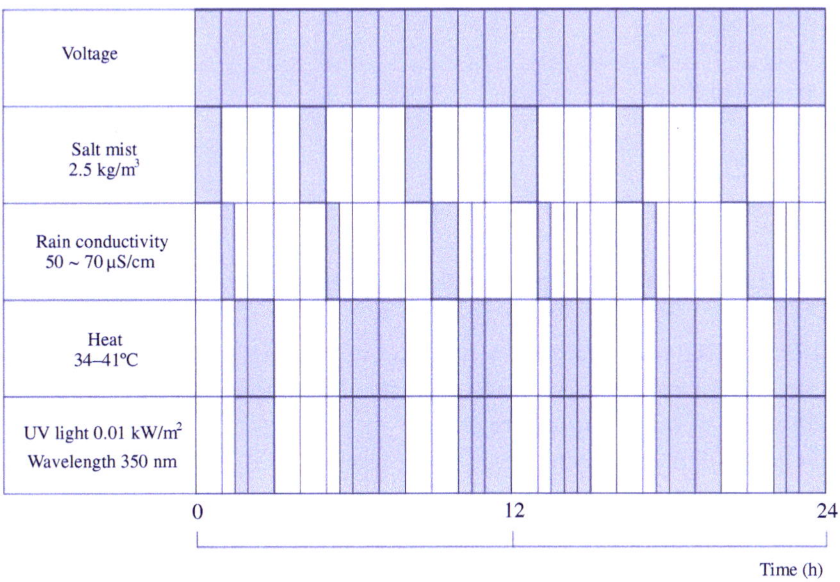

Figure 3.33 Winter cycle aging test procedure in EPRI summer/winter aging cycle procedure.

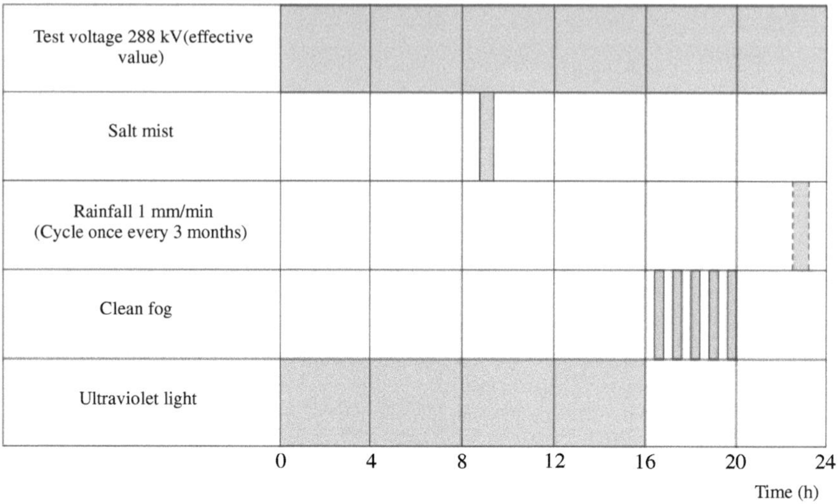

Figure 3.34 EPRI desert climate cycle aging test procedure.

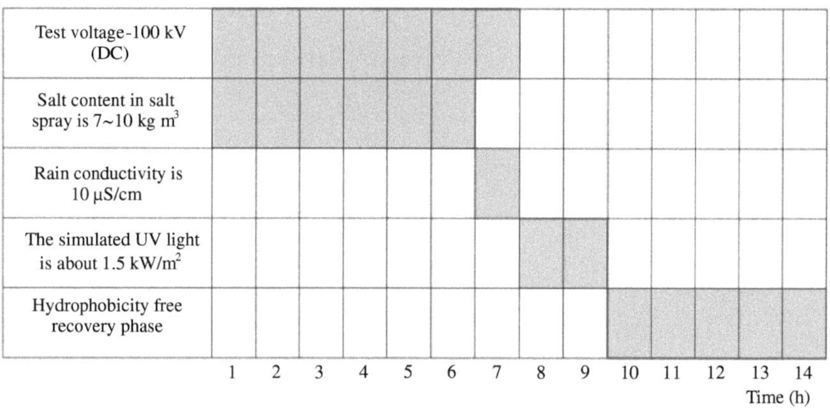

Figure 3.35 FGH 5000 h composite factor aging test procedure.

Both EPRI test procedures have more climate pertinence and have specific advantages, but also lead to greater limitations that have no application value in other climatic regions.

4. FGH 5000 h composite factor aging test

FGH 5000 h composite factor aging test is mainly applied to DC voltage; the test procedure is shown in Fig. 3.35. The test cycle period is 14 days, of which 6 days are with salt spray, 1 day is with rain, 2 days are with UV, and 5 days are hydrophobic recovery time without factors. Six days of salt spray period is too long, with the result that the ablation of the sample is more serious, and 2 days of UV irradiation time is short for the aging test. The overall result of the composite factor aging test is more similar to the IEC 1000 h salt spray method.

3.2.2.2 Carbon Fiber Composite Core Aluminum Conductor Test Conditions and Samples

Factors affecting the aging of carbon fiber composite core include including more UV in the sunlight, salt spray, acid rain, corona discharge and ozone, high temperature, and so on. Northern China Electric Power Research Institute draws lesson from artificial accelerated aging test methods widely used on organic composite insulators, by combining with the characteristics of the conductor, and designs the composite factor artificial accelerated aging test of carbon fiber composite core aluminum conductor, which lays a basis for safe operation and standard setting of the carbon fiber composite conductor and new organic material conductor that may appear in the future.

The test design refers to the GB/T 19519-2004 *Composite Insulators for AC Overhead Lines with Nominal Voltage Above 1000 V—Definitions, Test Methods and Acceptance Criteria,* and composite factor artificial accelerated aging test method for composite insulators recommended derives from the IEC method. The standard incorporates a variety of aging factors into a test project to simulate the actual operating environment relatively truly and appropriately increasingly, which has been generally accepted by testing agencies, manufacturing enterprises, and users of the composite insulators at home and abroad.

Carbon fiber composite core aluminum conductor test conditions are as follows: Taking 24 h as a cycle, the application of UV radiation, salt spray, rain, hot dry heat, and hot and humid conditions is cycled. The total test time is generally 5000 h; the cycle orders of each aging factor per 24 h are shown in Fig. 3.36. The differences from the composite insulator composite factor artificial accelerated aging test are the test design, the method of applying voltage, and the way of applying tension. In addition, the test system adds low-temperature test conditions, e.g., the minimum temperature of −40°C, which is mainly used to simulate the operating environment of the Northeast, Inner Mongolia, and other cold areas and can be used as a negotiated test condition.

The sample selects JLNR/F2B-240/45 type carbon fiber composite core heat-resistant aluminum alloy twisted conductor, and conductor parameters are shown in Table 3.4. The composite factor artificial

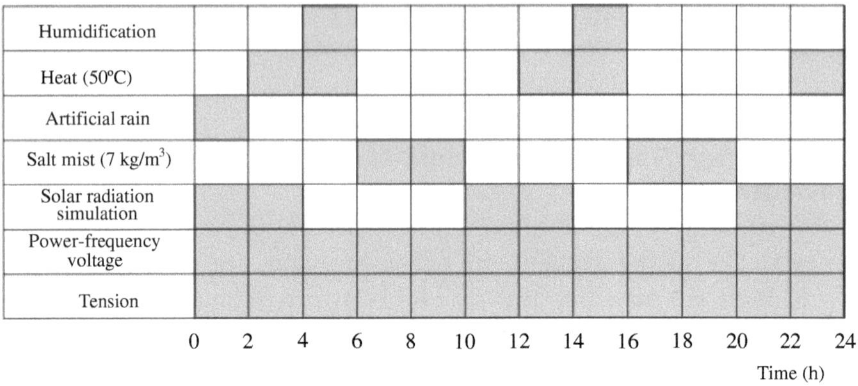

Figure 3.36 Cycle orders of each aging factor per 24 h.

Table 3.4 JLNR/F2B-240/45 conductor parameters

Composite core	Diameter	mm	7.5
	Cross-section	mm^2	44.18
Aluminum	The first layer (piece/diameter)	mm	9/3.70
	The second layer (piece/diameter)	mm	16/3.41
	Cross-section	mm^2	242.89
Comprehensive cross-section		mm^2	287.07
Conductor outer diameter		mm	21.7
DC resistance of stranded conductor		Ω/km	0.12354
Weight per unit length of stranded conductor		kg/km	738.5
Calculated tensile force		kN	120
Elastic modulus		GPa	71.45
Coefficient of linear expansion		$°C^{-1}$	14.9×10^{-6} (before the inflection point temperature) 2.0×10^{-6} (after the inflection point temperature)

accelerated aging test device is shown in Fig. 3.37. Carbon fiber composite core aluminum conductor is placed in the middle of the corona cage, corona cage is at ground potential, and power frequency voltage is applied to the conductor. The purpose of using a corona cage is to produce a

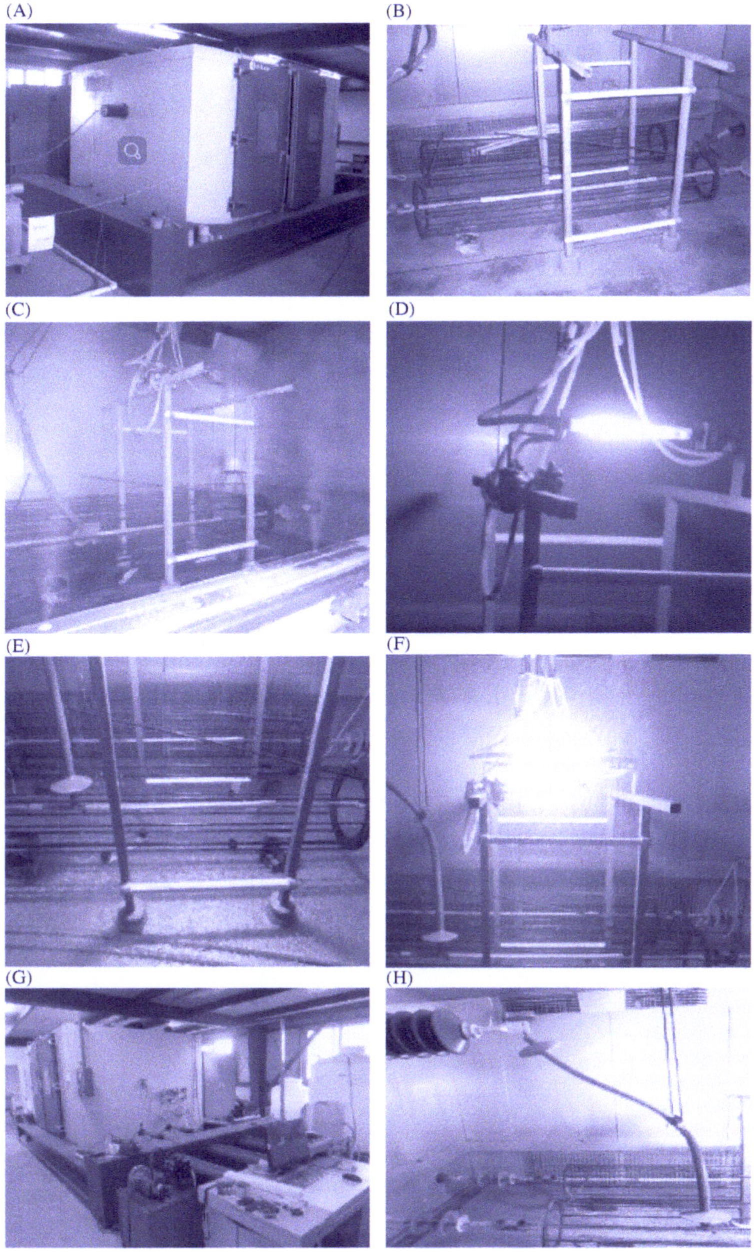

Figure 3.37 Composite factor artificial accelerated aging test device of carbon fiber composite core aluminum conductor (I).

(A) Composite factor aging box. (B) Carbon fiber composite core aluminum conductor sample; corona cage. Composite factor artificial accelerated aging test device of carbon fiber composite core aluminum conductor (II). (C) Salt mist. (D) Hernia lighting. (E) Rain. (F) Rain + hernia lighting. (G) Conductor tension applying means. (H) Applying frequency voltage.

corona discharge on the surface of the conductor and generate some ozone. The spacing between the conductor and corona cage is:

$$d = U \times 25/\left(110/1.732 \times 1.15\right) \times 1.5$$

wherein *U*—the frequency voltage applied to the conductor, kV; *d*—spacing between the conductor and corona cage, cm; 25—under 110 kV of power frequency voltage, the minimum safety distance between the conductor and tower, cm; 1.15–110 kV AC line, the ratio of highest voltage to rated voltage; 1.5—the safety factor, given by appropriately increasing the distance on the basis of the minimum clearance distance, considering the harsh conditions such as salt spray and rain in the test.

A certain amount of tension is applied to the conductor sample and the tension is 25% of the rated tensile force of the conductor. To facilitate the application of tension, both ends of the conductor are fitted with crimping strain clamps. Taking into account that the test mainly examines the antiaging properties of carbon fiber composite core, the steel anchor is crimped only on the carbon fiber composite core, but the aluminum tube is not crimped on the external aluminum strands, but the aluminum strands are still wrapped outside the carbon fiber composite core without bearing force, as shown in Fig. 3.38. In addition, a portion of carbon fiber composite core aluminum conductor sample and the carbon fiber composite core sample with aluminum strand peeled off are reserved and placed in an aging test box. During the test, parts of the sample are exposed to rain, salt spray, ultraviolet radiation, hot and dry heat, high

Figure 3.38 Installation of crimping strain clamp for carbon fiber composite core aluminum conductor.

temperature and humidity, and other test conditions, but do not take the tension and power frequency voltage.

3.2.2.3 Test Results

1. Mechanical and electronic microscope scanning test

   Composite factor artificial accelerated aging test lasts for 1000 h. After the aging test process, the appearance is inspected, followed by a tensile force test, three-point bending test, and crush test of carbon fiber composite core, assisted by electronic microscope scanning test, to make a more comprehensive identification analysis from macroscopic to microscopic on the conductor performance to evaluate the antiaging properties of carbon fiber composite core samples.

   a. Visual examination. The sample surface after the test is as follows:

      i. Carbon fiber composite core does not appear to have obvious corrosion, including corrosion caused by salt spray or corona.

      ii. Carbon fiber composite core does not appear to have cracking or fracture phenomenon.

      iii. Carbon fiber composite core surface appears lighter in color, as shown in Fig. 3.39(A).

      iv. Some corrosion spots appear on the surface of the carbon fiber composite core aluminum conductor aluminum strand and the color becomes dark, as shown in Fig. 3.39(B).

When carbon fiber composite core is exposed to light, high temperature, and other comprehensive aging factor conditions, the glass fiber

(A) (B)

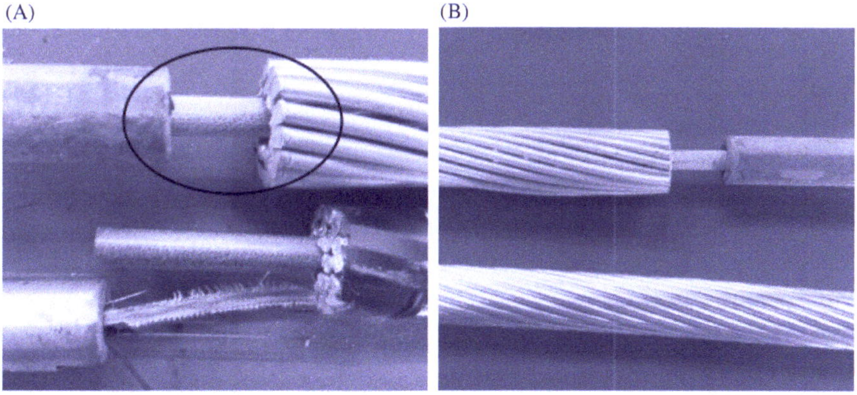

Figure 3.39 Sample surface after the test.
(A) Comparison of composite core before and after aging; (B) Comparison of aluminum strand before and after aging.

insulation layer impregnated by epoxy resin outside the composite core will become lighter in color, the result of which is within the range of the expected result of the test. When carbon fiber composite core aluminum conductor aluminum strands are exposed to salt spray and high temperature rain and other conditions, the surface appears to have a little oxidation corrosion, which is in line with normal operation, but is also within the expected results of the test.

b. Tensile force test. The tensile force test is divided into the carbon fiber composite core bar breaking test and aluminum strand breaking test. The comparison mainly focuses on the tensile forces of aluminum strand and composite core of the original sample, aging samples without tension and voltage, and aging samples with constant tension and voltage, to determine whether there is a clear drop in tension.

i. Carbon fiber composite core tensile force test (see Fig. 3.40). The original carbon fiber composite core is crimped with steel anchor and then undergoes directly tensile force test, tensile force of carbon fiber composite core is 120 kN. The original carbon fiber composite core aluminum conductor is crimped with supporting steel anchor, and placed into the aging test box, then is applied with a constant tension and power frequency voltage, tensile force test is performed after the aging test, tensile force of composite core is 118 kN; The original carbon fiber composite core aluminum conductor is placed into the aging test box, without applying tension and power frequency voltage, and is crimped with steel anchor after the aging test, tensile force of composite core is 128 kN.

ii. Aluminum strand tensile force test (see Fig. 3.41): The tensile forces of the three original aluminum strand samples are 1460, 1640, and 1550 N, respectively, and the average tensile force is 1550 N. After the aging test, tensile forces of three aluminum strand samples are 1550, 1580, and 1480 N respectively, and the average tensile force is 1537 N.

Aging test before and after the tensile force test shows that after 1000 h composite factor aging test for carbon fiber composite core aluminum conductor, tensile forces of carbon fiber composite core and aluminum strand are both not significantly reduced.

c. Three-point bending test. The three-point bending test of carbon fiber composite core mainly compares the three-point bending damage values of the original sample, the aging sample without tension

(A) (B) (C) (D)

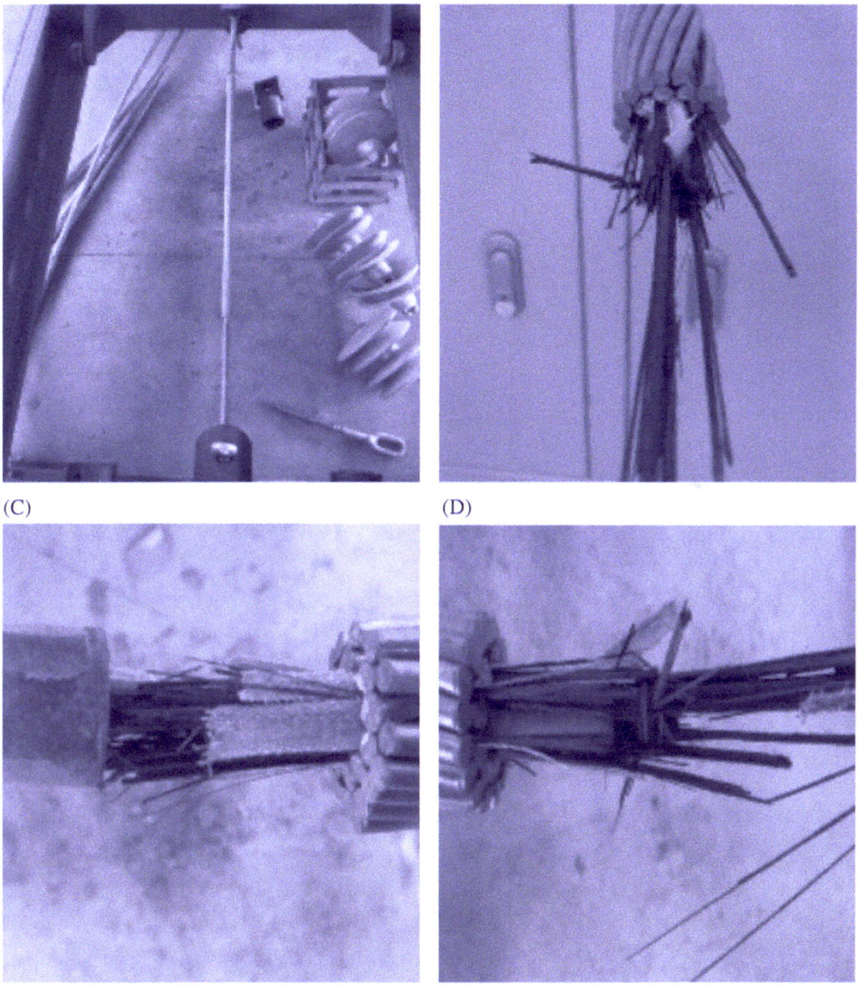

Figure 3.40 Carbon fiber composite core tensile force test and fracture.

and voltage, and the aging sample with constant tension and voltage to see whether they are significantly reduced. The length of the sample is 100 mm, the diameter D is 7.5 mm, and the test span is $6D$ (45 mm).

Three-point bending destructive forces of three original carbon fiber composite core samples are 2300, 2300, 2350 N, respectively; the average is greater than 2000 N. The original carbon fiber composite core aluminum conductor is crimped with a supporting steel anchor,

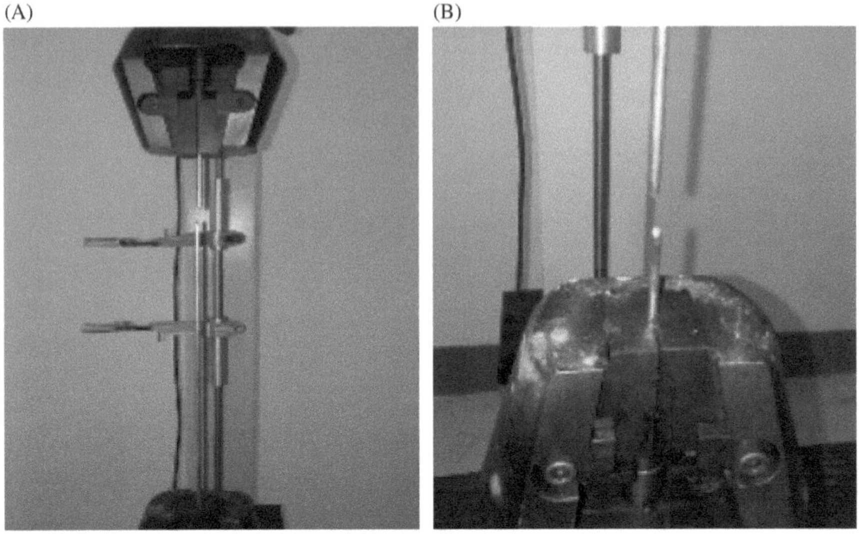

Figure 3.41 Aluminum strand tensile force test.

and placed into the aging test box, and constant tension and power frequency voltage are applied. After the aging test the carbon fiber composite core is cut into three samples that meet the size requirements for the three-point bending test; bending destructive forces are 2100, 2150, 2200 N, respectively, and the average value is greater than 2000 N. The original carbon fiber composite core aluminum conductor is put into the aging test box, without applying tension and power frequency voltage. After the aging test the carbon fiber composite core is cut into three samples that meet the size requirements for three-point bending test; bending destructive forces are 2140, 2100, and 2150 N respectively, and the average value is greater than 2000 N.

The three-point bending test of carbon fiber composite core before and after the aging test shows that after1000 h composite factor aging test for carbon fiber composite core aluminum conductor, three-point bending destructive forces of carbon fiber composite core are both not significantly reduced.

d. Flattening test. The flattening test of carbon fiber composite core mainly compares the flattening destructive force of the original sample, the aging sample without tension and voltage, and the aging sample with constant tension and voltage to see whether they are significantly reduced. The length of the sample is 100 mm, diameter D is 7.5 mm.

The flattening destructive forces of two original carbon fiber composite core samples are 15 and 16 kN, respectively, and average value is 15.5 kN. The original carbon fiber composite core aluminum conductor is crimped with a supporting steel anchor, and placed into the aging test box, applying constant tension and power frequency voltage. After the aging test the carbon fiber composite core is cut into two samples that meet the size requirements for the flattening destruction test; flattening forces are 12.4 and 12.7 kN respectively, and the average value is 12.5 kN. The original carbon fiber composite core aluminum conductor is put into the aging test box, without applying tension and power frequency voltage. After the aging test the carbon fiber composite core is cut into two samples that meet the size requirements for the flattening test; flattening forces are 12.5 and 12.5 kN respectively, and the average value is 12.5 kN.

The flattening destruction test of carbon fiber composite core before and after aging test shows that after 1000 h composite factor aging test for carbon fiber composite core aluminum conductor, flattening forces of carbon fiber composite core are not significantly reduced.

e. Scanning electron microscope (SEM) test. SEM tests mainly test, compare, and analyze the surface and fracture of carbon fiber composite core and aluminum strands before and after aging from a microscopic view, to determine the aging resistance of carbon fiber composite core aluminum conductor. An experimental program of the mechanical properties of carbon fiber composite core aluminum conductor and the electron microscope analysis are shown in Fig. 3.42.

Fracture necking of the original aluminum strand and aluminum strand after the aging test is obvious (see Figs. 3.43(A) and 3.44(A)); the fracture has mainly dimple fracture characteristics (see Figs. 3.43(B) and 3.44(B)).

There is no significant difference in the fracture characteristics of aluminum strands before and after aging test.

A portion of the fracture in the three-point bending failure test of the carbon fiber composite core is in the direction of parallel fiber, and the other portion of fracture is in the direction of vertical fiber. Fig. 3.45(A) and (B) shows the fracture characteristics in the direction of the vertical fiber. The fibers are broken, and the fiber arrangement characteristic is seen locally. A large amount of resin adheres to the fiber surface. Fig. 3.45 (C) shows the fracture characteristics in the direction of parallel fiber. Most of the fiber surface is cleaner and the resin is locally bonded.

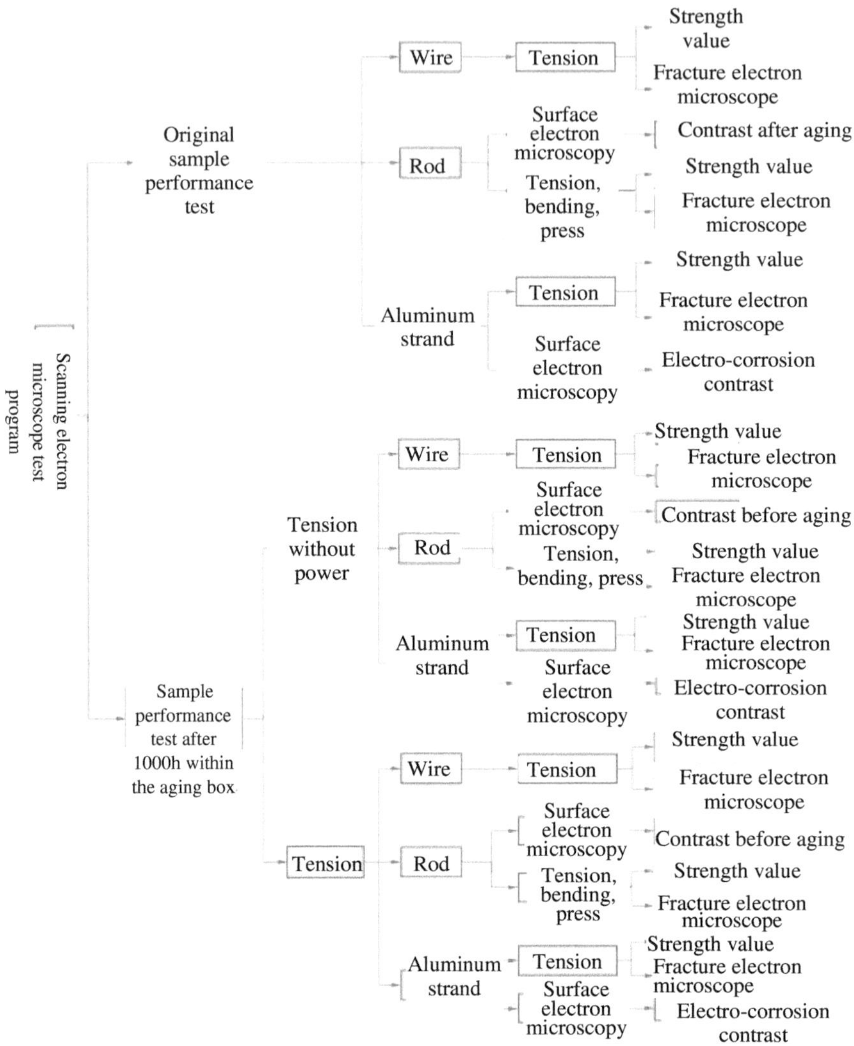

Figure 3.42 Experimental program view of conductor mechanical properties and electron microscopy analysis.

The surface of the outer glass fiber protective layer of the original carbon fiber composite core sample is more intact and continuous, as shown in Fig. 3.46. The surface of the glass fiber protective layer after aging has a certain degree of aging and the surface becomes rough, as shown in Fig. 3.47.

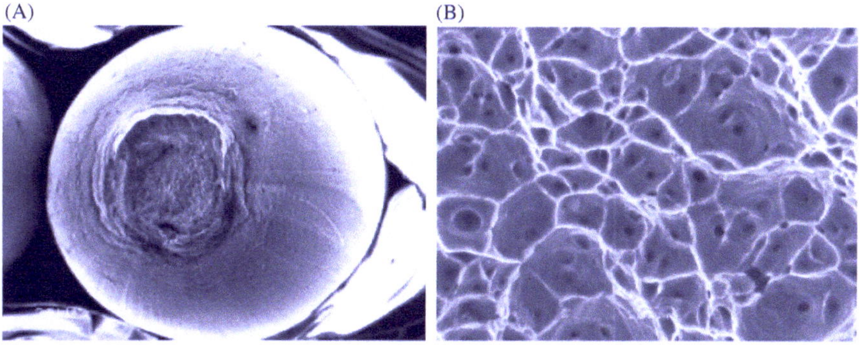

Figure 3.43 Original aluminum strand (tensile force of 1640).
(A) Aluminum strand fracture. (B) Dimple fracture characteristics.

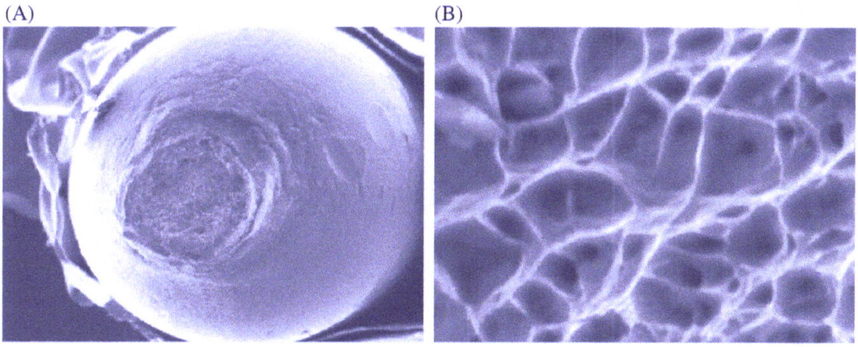

Figure 3.44 Aluminum strand after aging test (tensile force of 1450 N).
(A) Aluminum strand fracture. (B) Dimple characteristics.

SEM tests show that there is no significant difference in fracture characteristics between the original aluminum strand samples and the aluminum strand samples after the aging test. There is no significant difference in fracture characteristics of the three-point bending failure test between the original carbon fiber composite core sample and the carbon fiber composite core sample after the aging test. The surface of the outer glass fiber protective layer of the original carbon fiber composite core sample is more complete and continuous. After the aging test, the surface of the outer glass fiber protective layer has a certain degree of aging and the surface becomes rough without affecting the performance of the internal carbon fiber composite core.

(A)

(B)

(C)

Figure 3.45 Carbon fiber composite core three-point bending failure test fracture. (A) The fracture in the direction of vertical fiber of the original sample. (B) The fracture in the direction of vertical fiber of the sample after aging test. (C) The fracture in the direction of horizontal fiber of the original sample.

2. Differential scanning calorimetry (DSC), infrared spectroscopy, and energy spectrum analysis

   Carbon fiber composite core made by pultrusion or pultrusion and winding process of carbon fiber yarn and the appropriate resin matrix

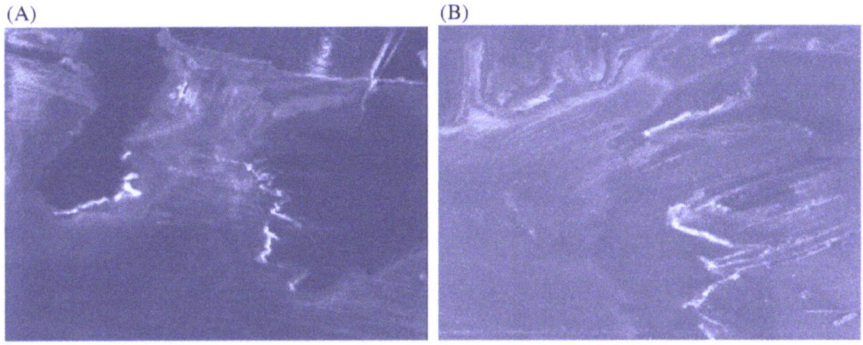

Figure 3.46 Surface of the outer glass fiber protective layer of the original carbon fiber composite core sample.

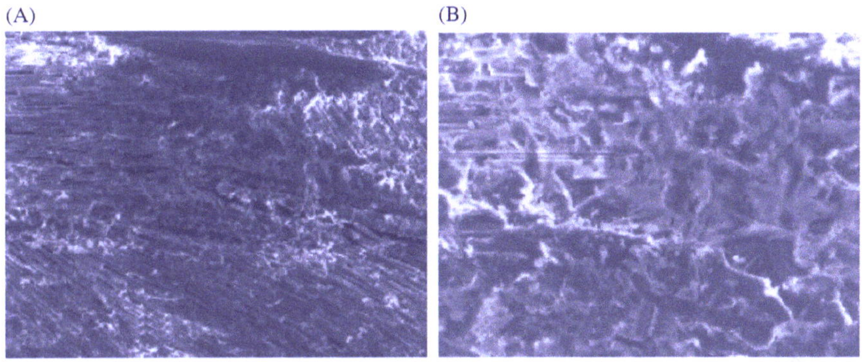

Figure 3.47 Surface of the outer glass fiber protective layer of the carbon fiber composite core sample after the aging test.

has superior corrosion resistance and resistance to atmospheric aging. Polymer material may undergo deterioration of the material properties during processing, application, and storage, such as yellowing, cracking, gloss loss, drop of impact strength, and other mechanical properties, thus affecting the normal use and service life of the polymer material, and this phenomenon is called the aging of polymer materials. The essence of aging is a chemical reaction that can be caused by a variety of factors such as heat, UV light, mechanical stress, high-energy radiation, electric fields, etc. As a result of aging, change in the structure of the polymer material and decrease or cross-linking of the relative molecular mass occur, eventually leading to the destruction of the material properties, and the material cannot be used.

The following DSC, infrared spectroscopy, and energy spectrum analysis examine the performance changes of carbon fiber composite core resin before and after the composite factor aging test.

a. DSC. DSC is based on the glass transition temperature, and the physical properties of the molecular polymers vary significantly with the degree of freedom of movement of the high molecular segments. The changes in heat capacity make the thermal analysis method an effective means for measuring polymeric material glass transition temperature. At present, the thermal analysis methods used for the determination of glass transition temperature are mainly differential thermal analysis and DSC. The DSC method is used in this experiment.

In Fig. 3.48, three curves 1–3 are DSC curves of original sample of the carbon fiber composite core, aging samples without tension and voltage, and aging samples with constant tension and voltage respectively. The curves show that within the range of 0–200°C, before and after the aging test, DSC of carbon fiber composite core has no significant change, that is, before and after the aging test, the resin matrix structure has no significant change.

b. Infrared spectroscopy analysis. Through the infrared spectrum and energy analysis, the changes of the outer surface of the carbon fiber

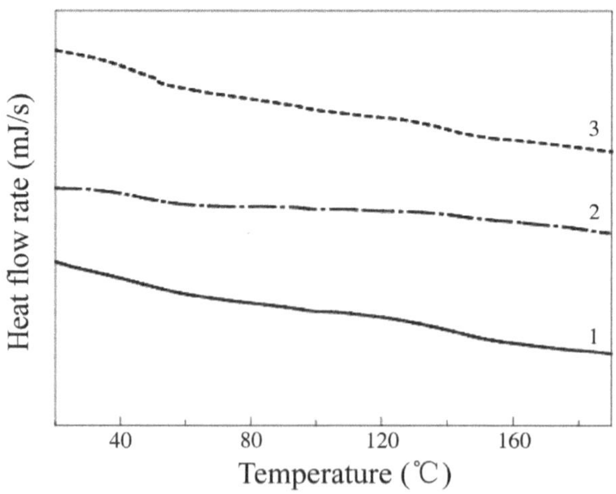

Figure 3.48 Differential scanning calorimetry curves.
1—Original sample; 2—aging sample without tension and voltage; 3—aging sample with constant tension and voltage.

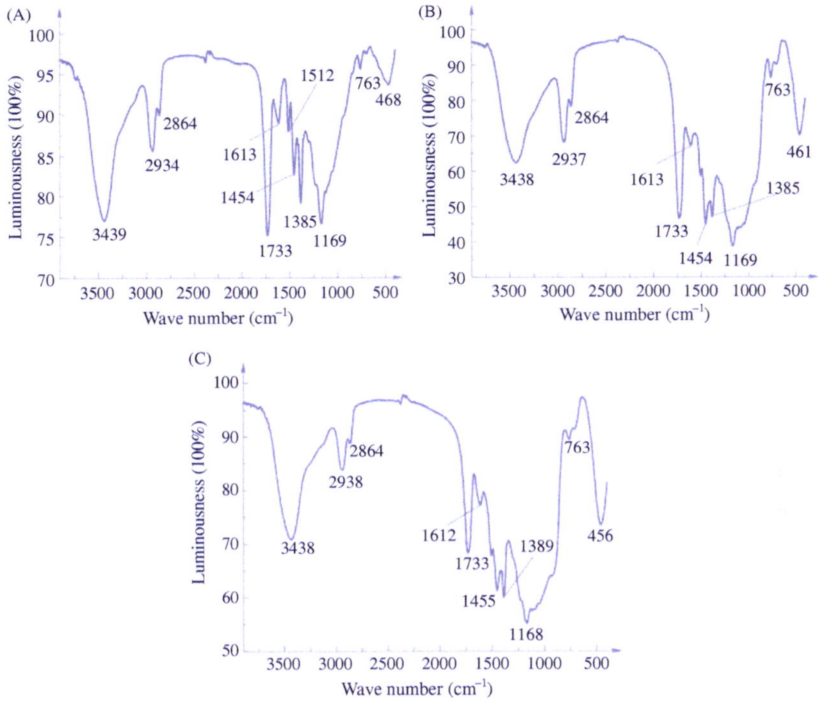

Figure 3.49 Infrared spectrum of carbon fiber composite core. (A) Composite core original sample. (B) Composite core aging sample without tension and voltage. (C) Composite core aging sample with tension and voltage.

composite core and the resin matrix structure before and after the aging test are determined.

Fig. 3.49(A) is the infrared spectrum of the original sample of the composite core, and Fig. 3.49(A) shows the in-plane rocking characteristic peak of CH2 at 763 cm^{-1}. At 1100–1247 cm^{-1} is a stretching absorption peak of ether C–O–C; at 1000–1100 cm^{-1} is a characteristic absorption peak of the fiber Si–O; at 1385 cm^{-1} is a symmetrical and antisymmetric deformation characteristic peak of the CH3; at 1512 and 1613 cm^{-1} is a characteristic absorption peak for the benzene ring; at 1733 cm^{-1} is a stretching vibration characteristic absorption peak of the ester group C = O; and at 2873–2934 cm^{-1} is a stretching vibration characteristic peak of the C–H of methyl and methylene; the characteristic absorption peak of –OH is at 3439 cm^{-1}, the spectrum coincides with the structure of the epoxy resin.

Fig. 3.49(B) and (C) are infrared spectra of yellowing surface of carbon fiber composite core aging samples without constant tension and voltage and carbon fiber composite core aging sample with the constant tension and voltage. Compared with Fig. 3.49(A), it can be seen that before and after the aging test, peak shape and peak position of the infrared absorption peak have no obvious change; that is, before and after the aging test, the carbon fiber composite core has no significant chemical changes, but at 1000–11,100 cm^{-1} the characteristic absorption peaks of Si–O increase, which is related to higher fiber content, aging, and oxidation of the surface of the material.

c. Energy spectrum analysis. The surface morphology and energy spectrum analysis results of carbon fiber composite core after aging test are shown in Table 3.5 and Fig. 3.50. Before and after the aging test, there is no significant difference in the energy spectrum elements of the epoxy resin impregnated glass fiber protective layer outside the carbon fiber composite core.

Table 3.5 Carbon fiber composite core surface energy spectrum analysis data after aging test

Element	Mass percentage (%)	Atomic percentage (%)
CK	65.89	72.19
OK	33.42	27.49
SiK	0.69	0.32
Total volume	100.00	

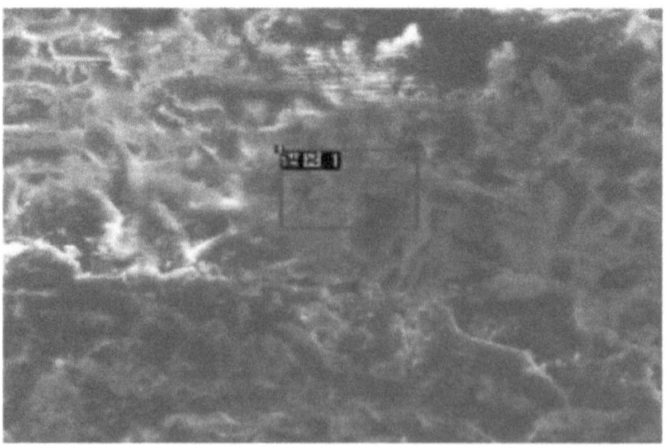

Figure 3.50 Composite core surface energy spectrum analysis after aging test.

3. Conductor galvanic corrosion analysis
   a. Galvanic corrosion theory. The carbon fibers composing the carbon fiber composite core are a kind of polycrystalline inorganic nonmetal material arranged along the fiber axis by incomplete graphite crystal. The structure and electrochemical properties of carbon fiber make its potential more positive, and the composition of the carbon fiber composite materials is inert in the generally corrosive environment. After carbon fiber is connected with the aluminum alloy point, galvanic corrosion can cause the corrosion rate of the aluminum alloy with the more negative electrode potential to be accelerated in the corrosive medium. Carbon fiber composite core aluminum conductor insulates the conductor from carbon fiber yarn with resin wrapped by the composite core outer layer and impregnated glass fiber protection wire, and during the operation the protective layer may appear to have a certain degree of aging phenomenon. This test tests and compares the galvanic current density between the aluminum strands and the carbon fiber composite core before and after the aging test to define whether the carbon fiber composite core and the aluminum strands can be used in contact with each other.
   b. Galvanic corrosion test program.
      i. Refer to the relevant standards and test the galvanic corrosion current between different materials according to the galvanic corrosion test program in Table 3.6.
      ii. The galvanic current–time curves of different galvanic pairs are measured according to the experimental scheme.
      iii. Using the galvanic current–time curve, the average galvanic current over 20 h is obtained by integral.
      iv. According to the actual area of the anode in the galvanic couple, the galvanic current density is found.
      v. The average galvanic current density, i_g, is calculated from the results of three parallel experiments.
      vi. The galvanic corrosion sensitivity is evaluated according to the amplitude of average galvanic current density.

According to the galvanic corrosion test scheme, the resin-impregnated glass fiber protective layer needs to be ground off one of the carbon fiber composite cores of the original sample. First, remove most of the protective layer and resin with a knife, and then grind by sandpaper.

Table 3.6 Galvanic corrosion test program

S/N	Galvanic pair		Test time (h)	Electrolyte
1	Carbon fiber composite core with protective layer unremoved	Aluminum alloy	20	3.5% NaCl solution
2	Carbon fiber composite core with protective layer removed	Aluminum alloy	20	3.5% NaCl solution
3	Carbon fiber composite core after aging test with tension and voltage	Aluminum alloy (after aging test)	20	3.5% NaCl solution
4	Carbon fiber composite core after aging test without tension and voltage	Aluminum alloy (after aging test)	20	3.5% NaCl solution

The carbon fiber composite core without the protective layer is shown in Fig. 3.51.

The galvanic dipole immersed in the brine requires the same surface area, and the length required for both can be found by the diameter of the aluminum strands and composite core, and the length of the composite core is set as 100 mm. Galvanic corrosion test is shown in Fig. 3.52.

For times involved in galvanic coupling tests of different materials and material conditions, please see Table 3.7.

The average galvanic current density between the carbon fiber composite core wrapped in the resin-impregnated glass fiber protective layer and the aluminum alloy is 0.0029 $\mu A/cm^2$, and the galvanic corrosion sensitive grade is grade A, that is, the carbon fiber composite core and the aluminum strand can be contacted for use. However, the average galvanic corrosion current density of composite carbon fiber composite core wrapped with resin-impregnated glass fiber protective layer and aluminum alloy galvanic coupling is 9.25 $\mu A/cm^2$, and the galvanic corrosion sensitive grade is D-class, that is, the carbon fiber composite core and the aluminum strand cannot be contacted for use. The average corrosion current densities between the carbon fiber composite core and the aluminum

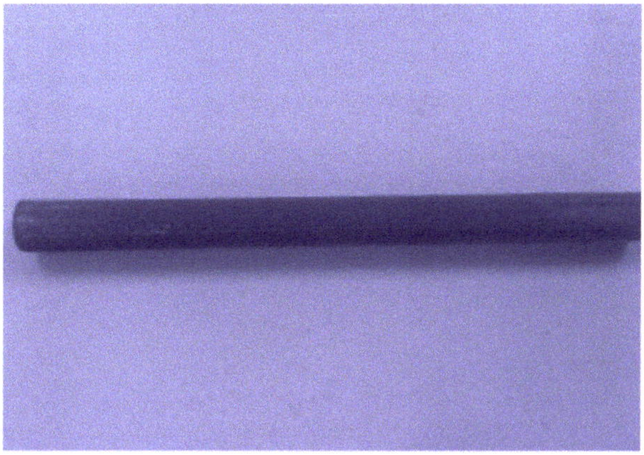

Figure 3.51 Carbon fiber composite core with the protective layer removed.

(A) (B)

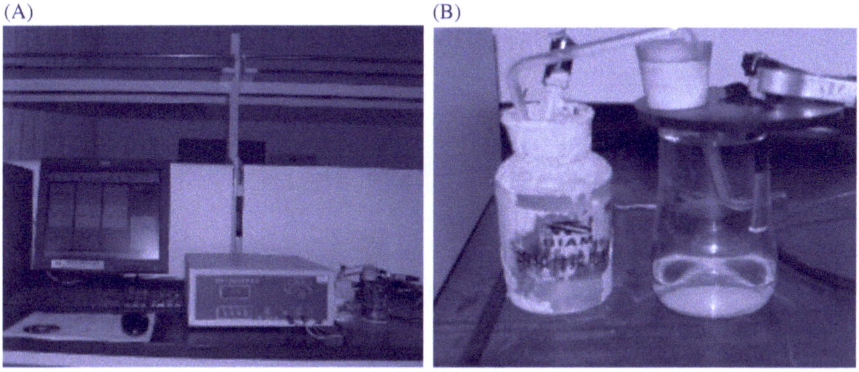

Figure 3.52 Galvanic corrosion test.

alloy galvanic coupling are 0.00116 and 0.00101 μA/cm^2, respectively after the aging test with constant tension and voltage and the aging test without tension and voltage. The galvanic corrosion sensitivity is Level A, that is, carbon fiber composite core and aluminum strand can still be contacted for use.

In summary, before and after the aging test, carbon fiber composite core aluminum conductor developed domestically does not appear to have the galvanic corrosion phenomenon, that is, the carbon fiber composite core and the aluminum strand can be contacted for use.

Table 3.7 Galvanic corrosion test data

S/N	Galvanic pair		Average galvanic current density ($\mu A/cm^2$)	Galvanic corrosion sensitive level
1	Carbon fiber composite core with protective layer unremoved	Aluminum alloy	0.00290	Grade A
2	Carbon fiber composite core with protective layer removed	Aluminum alloy	9.25	Grade D
3	Carbon fiber composite core after aging test with tension and voltage	Aluminum alloy (after aging test)	0.00116	Grade A
4	Carbon fiber composite core after aging test without tension and voltage	Aluminum alloy (after aging test)	0.00101	Grade A

3.3 MECHANICAL FATIGUE TEST

3.3.1 Mechanical Fatigue of Carbon Fiber Composite Core Aluminum Conductor

3.3.1.1 Mechanical Fatigue of Composite Material

The composite core of carbon fiber composite core aluminum conductor belongs to the organic composite material. Compared with metal materials, the composite material shows anisotropy in mechanical strength and rigidity, which makes its fatigue damage and failure mechanism more complicated than that of the isotropic metal material. There is no single damage mode in fatigue damage of composite materials, which is often a combination of matrix cracking, interfacial debonding, delamination, and fiber breakage. Therefore, it is relatively difficult to study the fatigue damage of composite materials. Even if macrocracking occurs in the composite material, the crack development does not occur in the same expected way as the metallic material, but in a constantly changing direction. The generation, expansion, and accumulation of fatigue damage may further aggravate the environment and stress corrosion of materials, accelerate the aging of materials, and cause further degradation of environmental resistance of materials and further loss of strength and stiffness, thereby reducing the service life of materials. Therefore, the fatigue resistance of composite materials has great influence on the reliability and safety of the material structure. Many scholars at home and abroad have conducted a great deal of research on the fatigue resistance of composite materials and their parts.

With the development of material science and manufacturing technology, composite materials have been widely used in power systems due to their advantages of light weight, high strength, corrosion resistance, and electrical properties. These uses include line composite insulators, composite interphase spacers, composite casing, composite towers, carbon fiber composite core aluminum conductors, and other products. However, the fatigue damage of composite materials appearing in power systems has also been increasing in recent years. For example, in December 2009, a compact line V-shaped composite insulator subjected to tension-compression/bending alternating stress in the long-term operation resulted in fatigue fracture. Expanded sampling shows that the same series of V-string composite insulators with faulty insulators also appears extremely similar to the phenomenon of fatigue damage; that is, fatigue damage has certain commonalities. Such failure shows that it is necessary to conduct an in-depth study on the mechanical fatigue resistance of organic composite materials for various types of power transmission and transformation equipment.

3.3.1.2 Mechanical Fatigue of Conductor

Material mechanics analysis and actual engineering tests show that stress of the overhead transmission conductor in the process can be divided into two types: static stress and dynamic stress.

1. Static stress

   Overhead transmission conductors are commonly fixed by embedded grip clamps. The static stress of the conductor at the connection of grip clamp mainly includes (1) tension directly from the conductor, (2) static bending stress caused by the bending of the conductor at the hanging point, and (3) radial extrusion stress formed at the clamp in the conductor connection support point by the wire tension, bending stress, and the internal stress of the conductor together.

   a. Tension: Tension in the overhead transmission conductor produces static stress in the conductor. Due to the stress of the conductor at different points in ACSR, the tension of conductor changes unevenly with the temperature. At a minimum temperature of −40°C, the internal stress of the conductor is up to

70 M–100 MPa, then the greater tension in the conductor makes self-vibration of the tension conductor increase, which can lead to serious vibration.

b. Static bending stress: The static bending stress of the overhead transmission conductor at the suspension clamp is related to the overhanging sag of the conductor and the tension of the conductor, and related to the flexibility of the conductor and the shape of the clamp.

c. Extrusion stress: The extrusion stress of overhead transmission conductors is an important part of the conductor force, which plays a significant role in the wearing of the conductor. Extrusion stresses are related to the linear properties of the suspension clamps and increase with the tension of the conductor and the holding force of the suspension clamp.

2. Dynamic stress

Overhead transmission conductors produce varying degrees of dynamic stress in response to icing, strong wind, and ambient temperature changes. According to the degree of vibration, dynamic stress can be divided into aeolian vibration, strong wind galloping, vibration induced by subspan oscillation, ice-shedding jump, and corona among split conductor subconductor.

a. Aeolian vibration: When a uniform breeze of 0.5–5 m/s vertically affects the overhead power transmission conductor, a whirlpool of air flow alternating up and down at a certain frequency will form at the leeward surface of the conductor. The vortex flow that appears will in turn subject the conductor to the impact of the same frequency up and down. The frequency f_w of the impact is related to wind speed v and the diameter D of the conductor, which is calculated according to $f_w = v/D$, with frequency range 5–150 Hz. The swirling of each point from the conductor is random, so the phase of the force acting on the conductor is also random along the length of the conductor. That is, it is not that a wind occurs accompanied by the conductor vibration. If the conductor vibrates at a certain frequency, and the frequency f_w and the phase of the impact force are within the range of the conductor frequency f_c, the separation of the vortex is controlled by the frequency of the conductor while the separation of vortices along the conductor forms a synchronization, and the aeolian vibration of the conductor starts. This phenomenon is often called the "synchronization effect." After the vibration starts, if the vibration frequency stays within the "synchronous" range, the lift acting on the conductor increases and the amplitude increases until the saturation amplitude is reached.

Aeolian vibration amplitude is small, and the vibration produces interactive stress. Only a small change occurs in the axial tension, and it will not have a greater impact on the static mechanical strength of overhead conductors. On the one hand, however, long-term aeolian vibration of the conductor may cause vibration hammer loosening, slip and fatigue damage of fittings, while on the other hand it may lead to mechanical fatigue of conductor material, and ultimately lead to broken strands or even broken conductors.

b. Galloping: When 5–15 m/s of wind acts on the conductor with asymmetrical shape, it can produce a pulsating force, prompting the conductor to form a large galloping, which is a result of aerodynamic instability—the vibration of overhead transmission conductor of a low frequency (about 0.1–4 Hz) and large amplitude (about 20–300 times the conductor diameter) caused by uniform icing and wind. During the galloping of the conductor, the vertical movement is often accompanied by movement in a twisting direction. The probability of galloping is relatively lower than aeolian vibration, but galloping energy is greater and the duration is longer, which is more harmful. There are two main hazards of galloping: mechanical and electrical. As for mechanical hazards, the large amplitude of galloping and accompanying twisting motion are a severe test for the strength fatigue resistance performance of the conductor and fittings. Severe galloping increases the bending stress of the conductor support point, and the tension will appear with alternating up and down in amplitude, which can lead to fatigue damage of the conductor, fittings, and accessories, and serious cases can cause disconnection and power failure. As for electrical hazards, due to the large amplitude of galloping, flashover between phases occurs easily, resulting in conductor burns and tripping accidents, thus affecting the reliability of line operation.

c. Subspan oscillation: Usually under the wind of 4–8 m/s, subconductor oscillation between the two separation rods in the overhead transmission line split conductor can cause subspan oscillation. Subspan oscillation is an unstable vibration of the subconductor at the leeward side induced by the wake flow of the subconductor at the upwind side, which is a form of vibration peculiar to split conductors. Subspan oscillation frequency is 1–3 Hz, and the amplitude thereof is up to several times to several tens of times that of the conductor diameter, which ranges between aeolian vibration and galloping. The main hazard involved in subspan oscillation is that it causes mechanical damage between the spacer and the conductor; in the area of split wire where there is severe vibration traces of fatigue damage to the bolt-type spacer can be found. However, the alternating bending stress originating from the vibration of the split conductor occurs at the junction of the wire and the spacer and has no effect on the static bending stress at the suspension clamp.

d. Ice shedding type jumping: Icing increases the overhead conductor vertical load, and the conductor produces elastic elongation. At the time of ice shedding, the icing on the overhead conductor breaks down and the elastic energy stored in the conductor is instantaneously released, forcing the overhead conductor to jump up and down. The ice shedding easily leads to the interphase short circuit between the conductors arranged up and down. At the same time, the tension of the conductor changes rapidly during the jump of the conductor, and the frictional damage easily occurs among the conductor, the suspension clamp, and the subconductor spacer.

During the operation of the overhead transmission line, the conductor is often subjected to the combined effect of static stress and dynamic stress. In particular, when the corridor of the line passes through the mountain ridge, the river tuyere, and other microtopography and micrometeorological areas, various complicated dynamic stresses may cause conductor fatigue damage, seriously affecting the safety and stability of operations and service life of the conductor.

3.3.1.3 Test Purpose

Compared with the traditional all-metal conductor, the application time of the carbon fiber composite core aluminum conductor is relatively short. The early mechanical fatigue test basically refers to the test method and evaluation criteria of ordinary ACSR (i.e., 3×10^7 aeolian vibration test). The aluminum strands of carbon fiber composite core aluminum conductor are similar to the conventional conductor, so the mechanical fatigue resistance is not the focus of attention. However, the carbon fiber composite core belongs to the organic composite material, and the performance is not the same as ordinary wire steel core; its long-term vibration fatigue resistance performance often becomes the concern of relevant units and personnel. Only the US CTC company conducts 10×10^7 vertical aeolian vibration test for its carbon fiber composite core aluminum conductor.

In light of the above, Northern China Electric Power Research Institute has simulated a variety of actual conditions of fatigue and vibration, and performs a 1×10^7 tension–tension fatigue test, 12×10^7 aeolian vibration test, 3×10^7 split conductor vertical vibration test, horizontal vibration test, and torsional vibration test for carbon fiber composite core heat-resistant aluminum alloy twisted conductor self-developed domestically, to conduct a more comprehensive and more rigorous examination of the carbon fiber composite core aluminum conductor in terms of the mechanical fatigue resistance performance in the long-term operating conditions to relieve the worries of the operating unit.

3.3.2 Carbon Fiber Composite Core Tension–Tension Stress Fatigue Test

The fatigue with the maximum stress and the minimum stress both being tensile stress is called tension–tension fatigue, and the test for examining materials performance to withstand tension–tension fatigue is called tension–tension stress fatigue test. The tension–tension stress fatigue test is widely used in various types of tensile components to examine dynamic load fatigue resistance performance.

Ordinary ACSR is entirely made of metal materials, including the outer layer of aluminum strand and steel core conductor, and mainly undergoes the tensile load in operation, and after 1×10^7 tension–tension stress fatigue test, the remaining strength of steel wire is only 30%–40% of its initial strength. As a mature product widely used in power systems of composite insulators, its inside contains a glass fiber rod impregnated by epoxy, and it mainly undergoes the tensile load in operation. The early test of mechanical fatigue performance shows that alternating tension–tension stress can cause fatigue damage to the core rod, and reduce the strength of the material and the connection strength of the fittings.

The composite core of carbon fiber composite core aluminum conductor is a carbon fiber drawing rod impregnated by high-temperature-resistant resin, and operates as a part undergoing the tensile load for the long term. Under the operating conditions of galloping, wind vibration, and subspan oscillation, the tension of the conductor changes, and the tension is also affected by many factors such as temperature, load change, and wind speed. That is, the carbon fiber composite core aluminum conductor will also withstand the action of alternating tension–tension stress during operation; especially in harsh operating environments, this action will be more prominent.

1. Test methods

A total of six samples of carbon fiber composite core aluminum conductors are selected for the test (JLNR/F2B-300/65 type; conductor parameters are shown in Table 2.2), of which three samples directly undergo static tensile load failure test. The remaining three samples are first subjected to the tension–tension stress fatigue test, then to the static tensile load failure test, and are compared with the samples that are not subjected to the tension–tension stress fatigue test. Tension–tension stress fatigue test conditions are shown in Table 3.8.

2. Testing data

    a. Static tensile load failure test: The three samples of carbon fiber composite cores are directly subjected to the static tensile load failure test, with an average breaking load of 180.3 kN. The test data are shown in Table 3.9.

    b. Residual strength after tension–tension stress fatigue test: The static tensile load failure test is performed on three carbon fiber composite core samples after tension–tension stress fatigue test, and it shows an average breaking load of 159.1 kN. The test data are shown in Table 3.10.

Table 3.8 Tension–tension stress fatigue test conditions

Temperature	Room temperature	Temperature	Room temperature
Humidity	60%	Minimum cyclic load (kN)	30.0
Media	Air	Cycle waveform	Sine wave
Sample sectional dimension (mm)	ϕ9.0	Cycle frequency (Hz)	5
Sample scale distance dimension (mm)	500	Cycle number	1.0×10^7
Minimum cyclic load (kN)	60.0		

Table 3.9 Results of static tensile load failure test

Sample serial number	Sample specification (mm)	Breaking load (kN)	Average breaking load (kN)
1	ϕ 9.0	178.7	180.3
2	ϕ 9.0	182.1	
3	ϕ 9.0	180.2	

Table 3.10 Residual strength test results of sample after fatigue test

Sample serial number	Sample specification (mm)	Maximum load (kN)	Minimum load (kN)	Cycle number	Breaking load (kN)	Average breaking load (kN)
1	ϕ 9.0	60.0	30.0	1.0×10^7	156.2	159.1
2	ϕ 9.0	60.0	30.0	1.0×10^7	161.4	
3	ϕ 9.0	60.0	30.0	1.0×10^7	159.8	

3. Test conclusion

a. As for the carbon fiber composite core samples, under the action of specified stress cycle and after a specified number of cycles (1.0×10^7) of tension–tension stress fatigue test, the surface morphology thereof is intact, and failure fracture does not occur.

b. The residual fracture load (159.1 kN) of the carbon fiber composite core sample can still reach 88.2% of fracture load (180.3 kN) of the sample without undergoing tension–tension stress fatigue test after a specified number of cycles (1.0×10^7) of tension–tension stress fatigue test under the action of specified stress cycles.

c. After 1.0×10^7 times of the tension–tension stress fatigue test, the residual strength of the carbon fiber composite core is still close to 90% of the initial strength. Under the same test conditions, the residual strength of steel is only 30%–40% of the initial strength. Experiments show that tension–tension stress fatigue performance of carbon fiber composite core is significantly better than the steel core.

3.3.3 Aeolian Vibration Fatigue Test of Conductor

1. Conductor aeolian vibration test system

As shown in Figs. 3.53 and 3.54, the conductor aeolian vibration test system can simulate the aeolian vibration condition during the operation

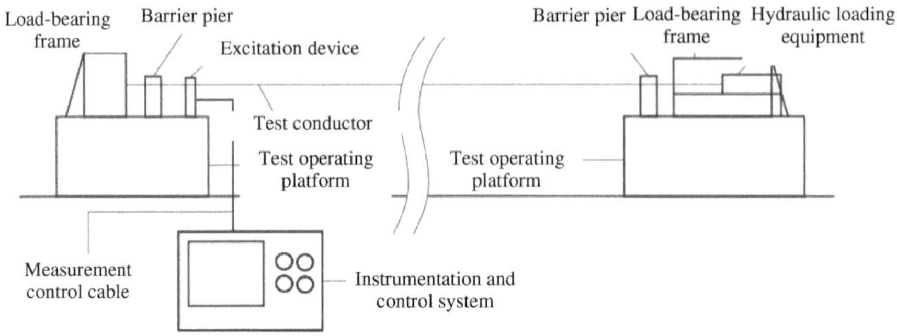

Figure 3.53 Schematic diagram of the conductor aeolian vibration test system.

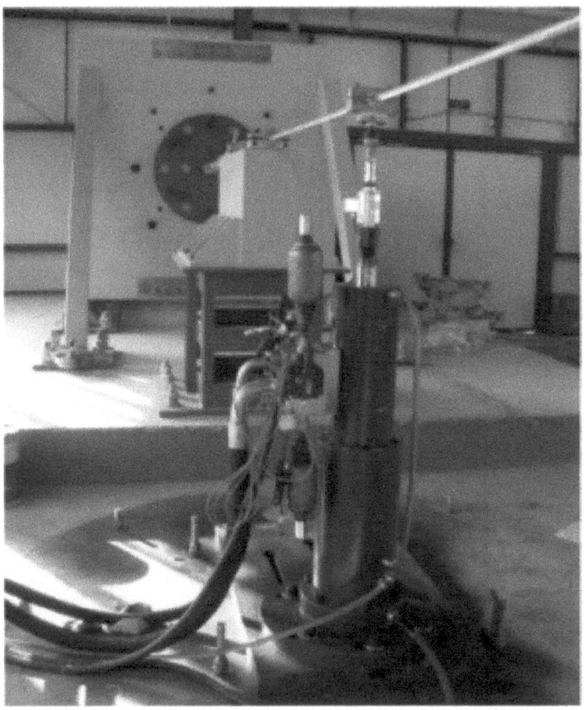

Figure 3.54 Conductor aeolian vibration test system terminal structure and excitation device system.

to investigate the vibration shape, shock-proof measures and vibration-proof level of the conductor during the aeolian vibration.

The test system consists of a tensioning device, traction device, barrier pier, excitation device, measurement control system, and other modules. Tensioning devices and traction devices are used to simulate operating conditions and apply tension to the conductor. After completion of conductor tension loading, the barrier pier clamps the conductor, such that the conductor vibration energy is isolated from tensioning and traction devices to avoid damage to the equipment during the conductor vibration. The measuring and control system is connected with the excitation platform for measuring the vibration signal and adjusting the output of the excitation platform to control the test parameters such as vibration frequency and amplitude.

2. Test methods

a. Test preparations: The sample is selected as JLNR/F2B-240/45 carbon fiber composite core heat-resistant aluminum alloy twisted conductor, and the conductor parameters are shown in Table 3.4. Both ends of the conductor are equipped with crimping type strain clamps, which are connected with the terminal structure and the tensioning mechanism at both ends of the test range, as shown in Fig. 3.55. The conductor is loaded through the tensioning mechanism until the conductor tension reaches 25% of the rated tensile force (30 kN). The ends of the conductor are clamped using the barrier pier clamp and the excitation device is connected to the conductor.

i. Loading and adjusting of excitation force. The excitation device is controlled through the measurement and control system to impose excitation force on the conductor, and the conductor vibration is adjusted in the amplitude and frequency to form a

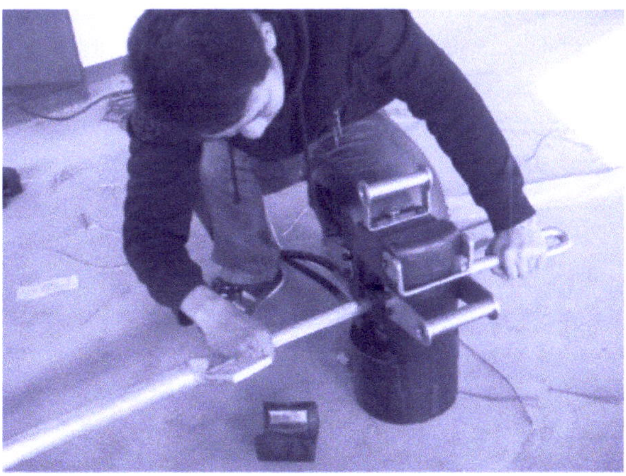

Figure 3.55 Conductor crimping strain clamps.

standing wave. The wavelength and amplitude of the conductor vibration are measured, the conductor vibration angle is calculated by the formula (3.1), and the vibration angle is further adjusted to the specified value of 30' by adjusting parameters such as amplitude and frequency.

$$\alpha = 60 \arctan \left(2\pi A/\lambda\right) \quad (3.1)$$

where A—amplitude (absolute amplitude), mm; λ—wave length, m; Test conditions. The final test parameters obtained are: frequency 39 Hz; vibration amplitude 10 mm; wave length 7.2 m; oscillation angle 30 degrees.

3. Test results
   a. The number of vibrations up to 3×10^7 times is recorded, the vibration test is suspended, and the excitation platform and barrier pier clamp outlet location are checked to find no breakage of silver strands.
   b. The vibration fatigue test is continued, the number of vibrations up to 6×10^7 times is recorded, the vibration test is suspended, and the excitation platform and barrier pier clamp outlet location are checked to find no breakage of aluminum strands.
   c. The vibration fatigue test is continued. The number of vibrations is recorded up to 7.2×10^7 times, and it is found that one aluminum strand at the barrier pier clamp outlet is broken. The vibration fatigue test is continued until the number of vibrations is up to 9×10^7 times and then suspended; then the excitation platform and barrier pier clamp outlet locations are checked, at which time it is found that still only the one aluminum strand at the barrier pier clamp outlet is broken.
   d. The vibration fatigue test is continued. The number of vibrations is recorded up to 10.3×10^7 times, and it is found that two aluminum strands at the barrier pier clamp outlet are broken (three aluminum strands in total are broken). The vibration fatigue test is continued until the number of vibrations is up to 12×10^7 times and then suspended. Then, the excitation platform and barrier pier clamp outlet location are checked, and it is found that still only the one aluminum strand at the barrier pier clamp outlet is broken. See Fig. 3.56.

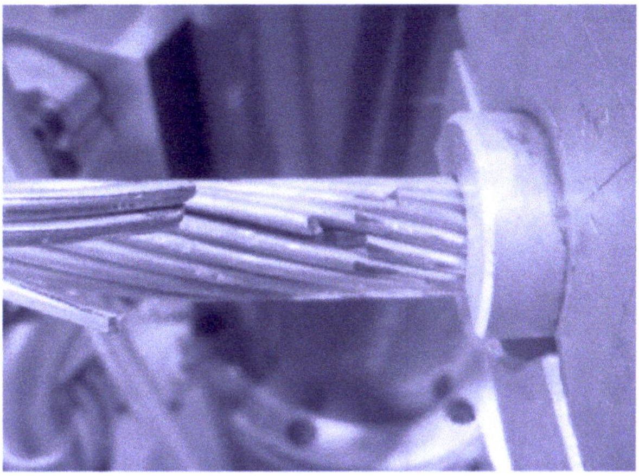

Figure 3.56 Broken conductor strand in a aeolian vibration simulation test.

e. Sample and aluminum strands are removed and the carbon fiber composite core is checked, and the carbon fiber composite core does not appear damaged.

When the number of vibration reaches 7.2×10^7 times the carbon fiber composite core aluminum conductor aluminum strand exhibits one broken strand, when the vibration number reaches 10.3×10^7 times the carbon fiber composite core aluminum conductor aluminum strand exhibits two broken strands; however at this time the test is far beyond ordinary ACSR general requirements (vibration number of 3×10^7 times). However, after 12×10^7 times of vibrations (which is four times the standard aeolian vibration number and greater than the number of vibration tests conducted by US CTC's carbon fiber composite core aluminum conductor, i.e., 10×10^7 times), the carbon fiber composite core remains intact, as shown in Fig. 3.57, so it is verified that the domestic carbon fiber composite core has excellent resistance performance to aeolian vibration fatigue.

3.3.4 Split Conductor Vibration Fatigue Test

3.3.4.1 Vibration Form of Split Conductor

Unlike aeolian vibration of a single conductor, the split conductors are connected to each other by subconductor spacers. During operation, the interconnection and confinement of the conductor and the subconductor spacer allow the combination of split conductor and spacers to form a variety of vibration forms under breeze, strong wind, galloping, icing, ice

Figure 3.57 Carbon fiber composite core inspection.

shedding, load changes, and other operating conditions, including vertical vibration, horizontal vibration, torsional vibration, online vibration, and so on. Under different types of vibration conditions, the interaction between the conductor and the subconductor spacer exerting restraining and limiting forces inevitably occurs, such as squeezing, friction, bending, and twisting, a combination thereof, etc. Alternating stress due to these actions can cause fatigue damage to both conductor and spacer.

This section describes the split conductor vibration fatigue test to evaluate fatigue resistance performance to the various types of vibration of the carbon fiber composite core aluminum conductor under the splitting arrangement conditions.

3.3.4.2 Split Conductor Vibration Fatigue Test System

The split conductor vibration fatigue test system is similar to the conductor aeolian vibration test system and consists of tensioning device, traction device, barrier pier, excitation device, measurement control system, and other modules; the schematic diagram of this system is shown in Fig. 3.58. The main difference is that the research object of the split conductor vibration fatigue test system is the conductor segment between two subconductor spacers, so the test span is relatively short (35 m), and the split conductor vibration fatigue test system focuses on the dynamic bending strain state of the wire at exit of the subconductor spacer, so the excitation device is installed in the central part of the span. The research object of the conductor aeolian vibration test system is the whole

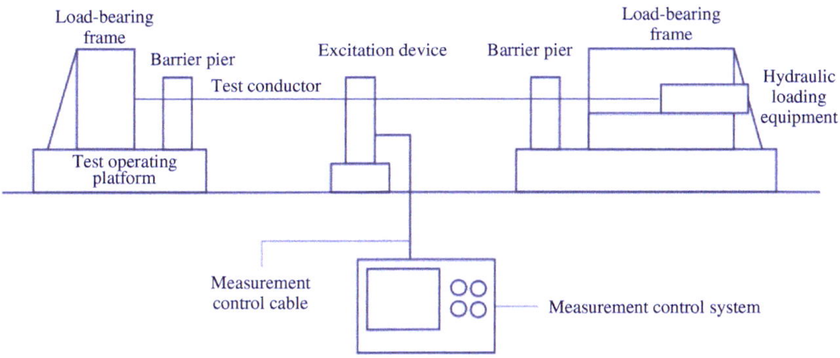

Figure 3.58 Split conductor vibration fatigue test system components diagram.

conductor section between two suspension clamps, and focuses on the dynamic bending strain of the conductor at the exit of the suspension clamp. Therefore, the excitation device is installed at one end of the span. In addition, by changing the vibration mode of the excitation device, the split conductor vibration fatigue test system can simulate a variety of actual operating conditions of the split conductor, including three vibration modes: vertical, horizontal, and torsional.

Tensioning devices and traction devices are used to simulate operating conditions and apply tension to the conductor. After completion of conductor tension loading, the barrier pier clamps to the conductor, such that the conductor vibration energy is isolated from tensioning and traction devices to avoid damage to the equipment during the conductor vibration. According to the test need, select excitation platform of appropriate vibration form to impose vibration on the conductor. The measuring and control system is connected with the excitation platform for measuring the vibration signal and adjusting the output of the excitation platform to control the test parameters such as vibration frequency and amplitude.

Different from the aeolian vibration fatigue test, the split conductor vibration fatigue test can install the subconductor spacer on the split conductor to check the cooperation and mutual influence between the conductor and the subconductor spacer under various vibration modes.

1. Vertical vibration

Vertical vibration refers to the up and down reciprocating vibration of the split conductor in the direction perpendicular to the conductor (that is, parallel to the plane of the subconductor spacer). The vibration mode is shown in Fig. 3.59. The excitation method of vertical vibration of the split conductor is similar to that of the aforementioned aeolian vibration

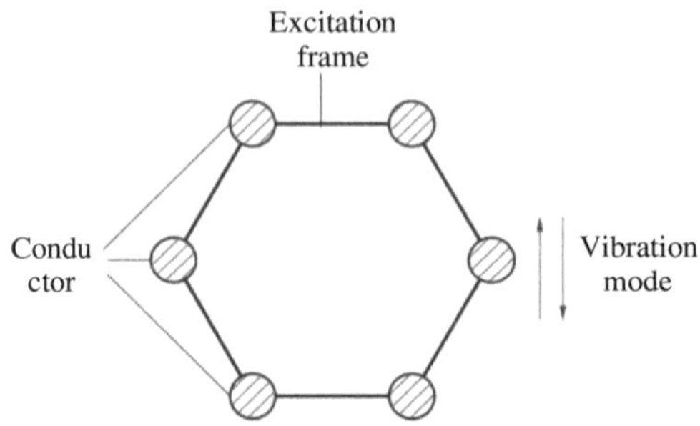

Figure 3.59 Schematic of conductor vertical vibration fatigue test vibration mode.

Figure 3.60 Conductor vertical vibration fatigue test excitation device.

test system. However, the excitation device synchronously vibrates multiple conductors simultaneously by excitation frame (see Fig. 3.60). The vertical vibration of the split conductor not only examines the vibration-resistant level of the conductor itself, but also investigates the effect of the interaction between the conductor and the subconductor spacer.

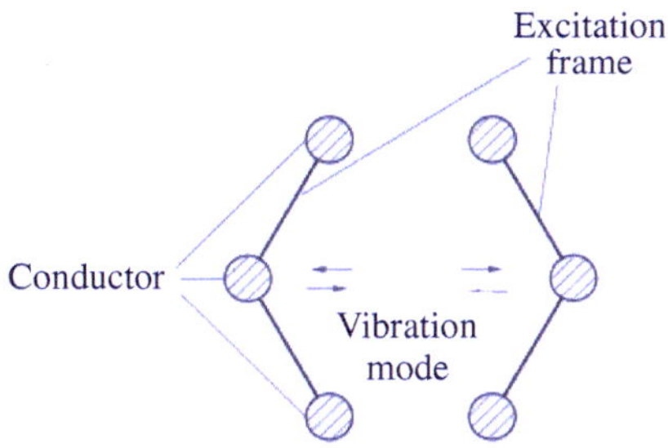

Figure 3.61 Schematic diagram of vibration mode of horizontal vibration fatigue test of conductor.

Figure 3.62 Horizontal vibration fatigue test excitation device of conductor.

2. Horizontal vibration

The horizontal vibration divides split conductors into left and right groups, and the relative reciprocating vibration in the horizontal direction is formed between the two groups. The schematic diagram of the vibration mode and the excitation device are shown in Figs. 3.61 and 3.62 respectively. Horizontal vibration mainly simulates operating conditions in which relative displacement is formed between the subconductors. Under conditions of split conductor oscillation in the subspan, the split conductor shows ice shedding jump out of sync, and the conductor undergoes electrodynamics force under heavy load. The relative vibrations of the subconductors during operation are generally not perfectly symmetrical, but for ease of design and laboratory implementation, this test system uses a simplified and perfectly symmetrical horizontal vibration pattern.

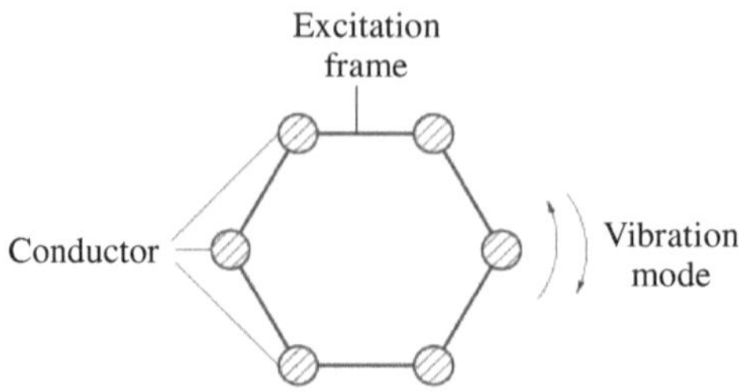

Figure 3.63 Schematic diagram of vibration mode of torsional vibration fatigue test of conductor.

Figure 3.64 Conductor torsional vibration fatigue test excitation device.

3. Torsional vibration

Torsional vibration refers to the overall twist reciprocating vibration of the split conductor in the plane parallel to the subconductor spacer. The schematic diagram of the vibration mode and the excitation device are shown in Figs. 3.63 and 3.64 respectively. This form of vibration occurs frequently when the multiple split conductors are exposed to the

Figure 3.65 5000 VShenbao second-line four-split conductor torsion phenomenon in April 2014.

wind, and is more violent as the gale wind or galloping occurs on the wiring or even causes the split conductors to twist together as shown in Fig. 3.65.

3.3.4.3 Test Methods

1. Test preparations

   The conductor sample is still selected as JLNR/F2B- 240/45 carbon fiber composite core heat-resistant aluminum alloy twisted conductor, and the conductor parameters are shown in Table 3.4.

   Six conductors equal in length make up six split conductors, a crimping strain clamp is installed at both ends of each conductor, and both ends are connected to a terminal structure and tension mechanism at both ends of the test section. The conductor is loaded through the tensioning mechanism and the tension on each wire conductor is adjusted using a tension adjuster until it reaches the balance of six conductors and tension reaches 25% of the rated tensile force (30 kN). Clamp both ends of the conductor with the barrier pier, connect the excitation device to the conductor, and install the subconductor spacer at the conductor that is 5 m far away from the excitation platform, as shown in Fig. 3.66.

2. Loading and adjusting of excitation force

   The excitation device is controlled through the measurement and control system to impose excitation force on the conductor, and the

Figure 3.66 Split conductor being equipped with subconductor spacers.

conductor vibration is adjusted in the amplitude and frequency to reach a stable status.

3. Test conditions
   a. Vertical vibration: Frequency, 36 Hz; vibration amplitude, 8 mm.
   b. Horizontal vibration: Frequency, 10 Hz; vibration amplitude, 5 mm.
   c. Torsional vibration: Frequency, 10 Hz; vibration amplitude, 4 degrees.

3.3.4.4 Test Results

1. Vertical vibration: The number of vertical vibrations is recorded until it reaches 3×10^7 times, the vibration test is stopped, the excitation device, barrier pier, and clamp exit position of the subconductor spacer are checked, and no conductor aluminum strand breakage is found. The clamp of the subconductor spacer is further opened to check the conductor, where it is found that the aluminum strands of the clamp holding part are intact; see Fig. 3.67. Aluminum strands of the subconductor spacer holding part are peeled off, and the carbon fiber composite core and the composite core are checked and found to be intact, as shown in Fig. 3.68.
2. Horizontal vibration: The number of horizontal vibrations are recorded until it reaches 3×10^7 times, the vibration test is stopped, and the excitation device, barrier pier, and clamp exit position of subconductor spacer are checked, at which time no conductor aluminum

Figure 3.67 Open subconductor spacer clamp for check.

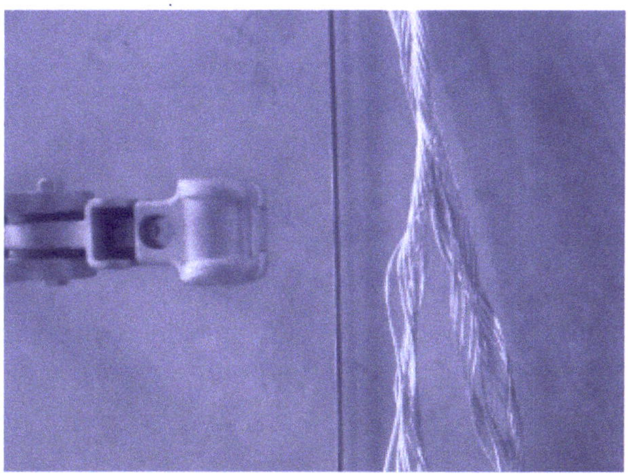

Figure 3.68 Anatomic carbon fiber composite core aluminum conductor for inspection.

strand breakage is found. The clamp of the subconductor spacer is then opened to check the conductor, and the aluminum strands of the clamp holding part are intact. Aluminum strands of the subconductor spacer holding part are peeled off, and the carbon fiber composite core and the composite core are found to be intact.

3. Torsional vibration: The number of torsional vibrations are recorded until it reaches 3×10^7 times, the vibration test is stopped, and the excitation device, barrier pier, and clamp exit position of subconductor spacer are checked, at which time no conductor aluminum strand

breakage is found. The clamp of the subconductor spacer is opened to check the conductor, and the aluminum strands of the clamp holding part are intact. Aluminum strands of the subconductor spacer holding part are peeled off, and the carbon fiber composite core and the composite core are intact.

3.4 SUMMARY

Comprehensive test of carbon fiber composite core aluminum conductor shows:

1. The carbon fiber composite core aluminum conductor developed domestically meets the requirements of conventional conductor type tests, and in terms of mechanical strength, mass, sag, and other indicators is significantly better than the conventional conductor.
2. The carbon fiber composite core developed domestically is significantly better than the US CTC products in terms of high temperature tensile force and anticrush performance.
3. Aluminum strand of carbon fiber composite core heat-resistant aluminum alloy twisted conductor has high mechanical strength, and is not easy to deform or damage during construction, operation, and maintenance. Carbon fiber composite core soft aluminum conductor has high conductivity and the advantage of low operating loss. Aluminum strand of carbon fiber composite core duralumin twisted conductor has both advantages of mechanical strength and transmission loss, which is one of the development directions of the carbon fiber composite core aluminum conductor.
4. The composite factor artificial accelerated aging test method is designed for carbon fiber composite core aluminum conductor. The excellent aging resistance of the carbon fiber composite core aluminum conductor developed domestically is verified. After 1000 h of composite factor artificial accelerated aging test, mechanical tests, and electron microscopy analysis, it shows that composite core sample has no significant loss of mechanical properties. DSC, infrared spectroscopy, and energy spectrum analysis show that the composite core resin matrix does not change significantly. Galvanic corrosion tests show that the carbon fiber composite core aluminum conductor does not exhibit galvanic corrosion, and the aluminum strand and composite core can be contacted.

5. The 12×10^7 aeolian vibration fatigue test, 3×10^7 split-conductor vertical vibration fatigue test, 3×10^7 split-conductor horizontal vibration fatigue test, and 3×10^7 split-conductor torsional vibration fatigue test are conducted for the first time on carbon fiber composite core aluminum conductor. A comprehensive study of the carbon fiber composite core aluminum conductor mechanical fatigue properties is performed. Domestic carbon fiber composite core aluminum conductor has mechanical fatigue performance no less than or better than the conventional conductor, and can be used in overhead transmission lines for the long term.

CHAPTER 4

Matched Fittings, Construction, and Operating Maintenance of Carbon Fiber Composite Core Aluminum Conductor

Contents

4.1 CONNECTING FITTINGS FOR NORMAL CONDUCTORS

Connecting fittings (including strain clamp, straight connecting pipe, etc.) are key elements for connections between conductor and tower as well as conductor and conductor and the emphasis of conductor in the process of construction and maintenance. Strain clamps connect the conductor to

Engineering Energy Aluminum Conductor Composite Core (ACCC) and its Application
DOI: https://doi.org/10.1016/B978-0-12-815611-7.00004-1

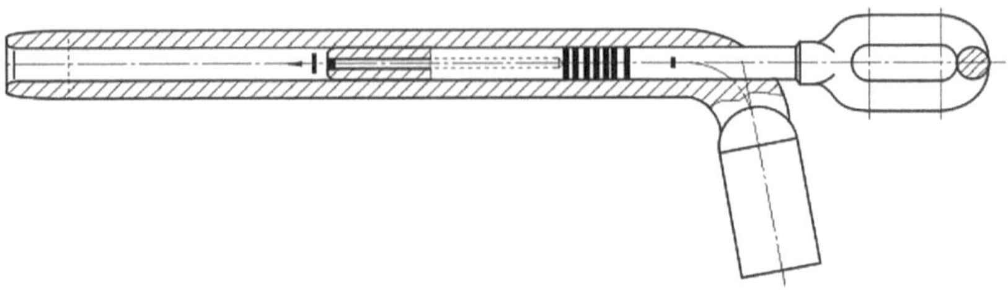

Figure 4.1 Structural diagram of compression-type strain clamp.

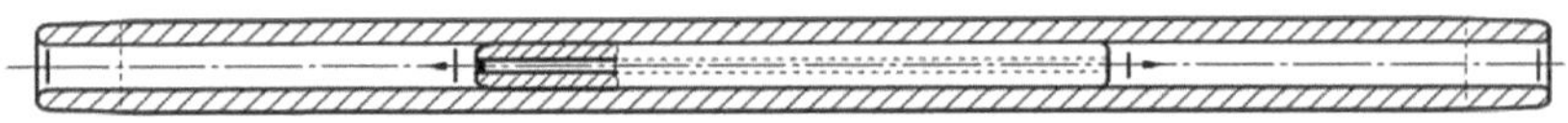

Figure 4.2 Compression-type straight connecting pipe.

the strain insulator string, and further connect it to the tower cross arm for fixing the conductor and bearing conductor tension as well as providing current-carrying capability. Strain clamps generally cover screw-type cable clamps, compression-type (pressure-type) strain clamps, and wedge-shaped strain clamps. Straight connecting pipe is used for conductor continuity or repatching, bearing conductor tension and providing current-carrying capability.

Compression-type strain clamps have a simple structure and are easy to install. As shown in Fig. 4.1, aluminum conductor steel reinforced (ACSR) goes through the aluminum tube and is inserted into the steel anchor. Crimping between steel core and steel anchor as well as between aluminum tube and aluminum strand is implemented with installation easily ensured; as a result, most high-strength normal conductors use a compression-type strain clamp.

Straight connecting pipe structure of a normal conductor is shown in Fig. 4.2. Its installation is similar to that of a strain clamp. The conductor connection is inserted into the straight connecting pipe from both ends, of which the steel core stretches into steel pipe internal docking or adopts loose strand bonding, and then crimping is conducted outside the steel pipe to connect the steel pipe and steel core, which is mainly used for bearing conductor tension. The crimping between the aluminum strand of the conductor and aluminum tube installed on it is conducted, and is mainly used for flow-through of the conductor and bearing partial conductor tension.

The above connecting fittings have a simple structure, and they can well meet the operating requirements of normal conductors. Especially, the connecting fittings are only constituted by internal steel pipe and external aluminum tube, so minimum length and outer diameter can be realized in terms of meeting mechanical strength and flow capacity, and it is easy for connecting fittings to accomplish continuous tension stringing with straight connecting pipe by tackle for paying-off and connecting fittings have been widely used in related fields. However, the above strain clamp and straight connecting pipe are designed in accordance with mechanical and electrical performance of normal conductors (ACSR, etc.); as a result, they fail to meet the related requirements of carbon fiber composite core aluminum conductor.

4.2 WEDGE JOINT CONNECTING FITTINGS AND EXISTING PROBLEMS

The major difference between carbon fiber composite core aluminum conductor and normal ACSR is that the steel core is replaced by carbon fiber reinforced composite core, and such difference leads to major differences between the installation of connecting fittings of carbon fiber composite core aluminum conductor and that of ordinary steel-cored aluminum stranded wire. The earliest and most widely used matched connecting fittings of carbon fiber composite core aluminum conductor are the wedge joint connecting fittings made by the US CTC Company, including wedged joint strain clamps and straight connecting pipes (refer to Fig. 4.3).

The main function of wedge joint connecting fittings is to realize fixation or clamping of carbon fiber reinforced composite core by wedge structure in the fittings. Strain clamps consist of five parts: the inner cone

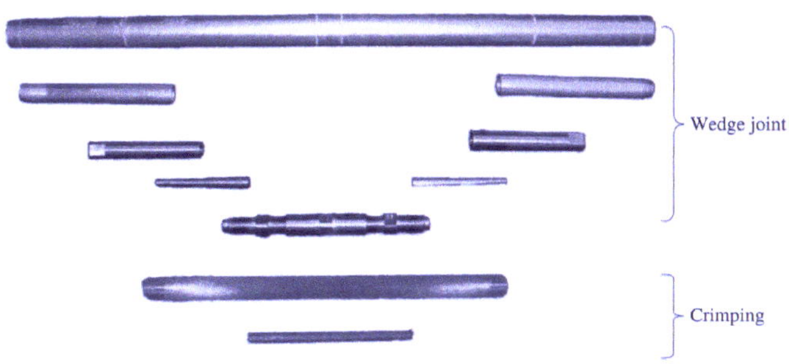

Figure 4.3 Wedge joint straight connecting pipe and clamped straight connecting pipe.

Figure 4.4 Inner cone hoop of wedge joint connecting fittings.

hoop (refer to Fig. 4.4), outer cone hoop, steel anchor, aluminum liner tube, and aluminum tube. The wedge structure formed by the inner and outer cone hoops enclasps the carbon fiber reinforced composite core and bears conductor tension based on the connection between steel anchor and tower. Aluminum strand is connected based on crimping, and crimping between aluminum strand and aluminum liner tube and aluminum tube installed on it is conducted to carry out current transmission. The principle of straight connecting pipe for carbon fiber composite core aluminum conductor is similar to that of strain clamp, and two sets of back-to-back inner cone hoops and outer cone hoops form the reverse wedge structure, grasping ends of carbon fiber reinforced composite core to be connected on both sides; crimping between aluminum strand and aluminum liner tube and aluminum tube is conducted for carrying out current transmission.

Although the above wedge joint connecting fittings are characterized by large grip strength, no large construction machines are needed for installation, etc., and there are many wedge structure components of fittings, resulting in expensive matched wedge joint fittings of the carbon fiber composite core aluminum conductor. The price of foreign product is more than 10 times that of matched fittings of normal ACSR, and the price of developed domestic wedge joint product is found to be more than five times that of normal matched fittings. In addition, because the straight connecting fittings adopt back-to-back wedge connection formed by the inner cone hoop (refer to Fig. 4.4) and outer cone hoop, the length and diameter of straight connecting pipe greatly increases in comparison with that of the straight connecting pipe of a normal conductor. The length of normal straight connecting pipe is generally not more than 80 cm, while the length of wedge joint straight connecting pipe for carbon fiber composite core aluminum conductor exceeds 170 cm. Although related units in China make an improvement in the length of wedge joint straight connecting pipe and shorten the total length, the diameter of wedge joint straight connecting pipe is difficult to reduce (unless flow capacity or mechanical strength of straight connecting pipe are sacrificed).

Because wedge joint straight connecting pipe is too long and has too large an outer diameter, it is easy to break it off or to damage the carbon fiber reinforced composite core of the conductor at the time of passing tackle, that is to say that it is difficult for the wedge joint straight connecting pipe (including steel armor for protection temporarily installed at the time of setting out) to go through the tackle for paying-off. In other words, the wedge joint straight connecting pipe makes it difficult for carbon fiber composite core aluminum conductor to realize continuous tension stringing in the course of track construction. Segmented paying-off (one segment not exceeding 5 km) is adopted at the time of construction, which brings great difficult to long-distance track construction with frequent transitions in the construction process, greatly increasing workload, work difficulty and construction cost. When construction occurs in the mountains, with bad environment and complicated terrain, if there is no appropriate traction field or tension field, it is hard to realize tension stringing; while high cost will arise due to expropriation of tension field and traction field in areas with dense population. Therefore, failure in continuous tension stringing becomes one of the bottlenecks regarding carbon fiber composite core aluminum conductor applied by long-distance track.

4.3 CRIMPING CONNECTING FITTINGS AND TENSION STRINGING

4.3.1 Development of Crimping Connecting Fittings

To promote popularization and application of carbon fiber composite core aluminum conductor and give full play to its technological

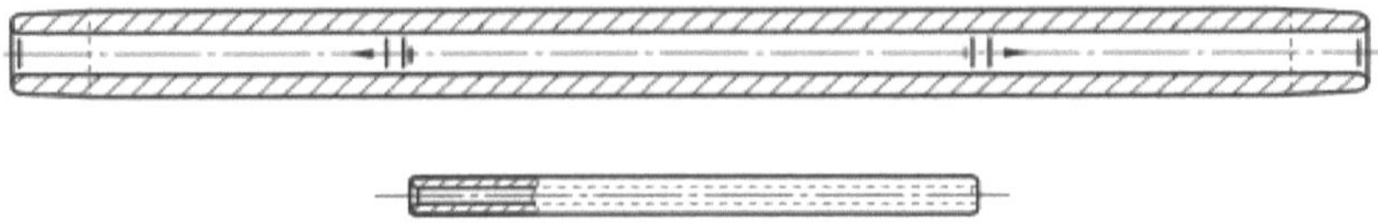

Figure 4.5 Crimping straight connecting pipe used for carbon fiber composite core aluminum conductor.

superiority in the power grid, after studying the crimping connection between crimping matched fittings of normal ACSR with composite insulator core bar of the track and fittings on the end, the Northern China Electric Power Research Institute has completed independent research and development of new matched fittings mainly characterized by crimping that are applicable to carbon fiber composite core aluminum conductor, including strain clamp and straight connecting pipe (refer to Fig. 4.5). Mechanical performance and electrical performance requirements of fittings are ensured, and the length and outer diameter of connecting pipe are significantly shortened at the same time. Taking JLNR/F2B-240/45 conductor as an example, a length of straight connecting pipe is shortened to 70 cm from 170 cm, which is the length of the wedge joint pipe, solving the problem of long-distance continuous tension stringing of carbon fiber composite core aluminum conductor in the construction process. What's more, compared with matched fittings based on foreign technology, the price of the above fittings is greatly reduced to only 10% of that of the foreign wedge joint product. Considering that continuous tension stringing fails to be realized, which causes frequent transitions resulting in substantial increases in construction and compensation costs, the economic benefit will be more considerable.

To ensure safety and reliability of matched fittings of crimping carbon fiber composite core aluminum conductor, a series of tests against crimping fittings have been conducted, including the crimping parameter test, static normal temperature grip strength test, high-temperature grip strength test, test of dynamic fatigue vibration of 30 million times, test of passing the tackle, and temperature rise test. The samples are JLNR/F2B-240/45 carbon fiber reinforced composite core heat-resisting aluminum-alloy conductor and matched crimping strain clamp and/or splicing sleeves, and calculated tensile force of JLNR/F2B-240/45 conductor is 120 kN.

4.3.1.1 Crimping Parameter Test

4.3.1.1.1 Test Device

1. Conductor crimping machine. The EP-200W conductor crimping machine is composed by hydraulic pump and crimping platform, and it is equipped with a full set of crimping molds. Replacement of crimping molds on the basis of strained tubes and straight connecting pipe with different dimensions is shown in Fig. 4.6. The pressure gauge of the hydraulic machine should be replaced with a smaller pressure gauge that reflects the actual pressure scope of the crimping carbon fiber reinforced composite core.
2. 1000 kN horizontal tension testing machine. Used for testing tensile failing load of conductor and connecting fittings.

4.3.1.1.2 Test Procedure and Result

Straight connecting pipe and strain clamp share the same crimping part, that is to say that their crimping effect is consistent, and as a result, the strain clamp is used as a test fitting for the purpose of facilitating the strain relief test and saving costs. JLNR/F2B-240/45 conductor is used for the test with the diameter of carbon fiber reinforced composite core being 7.5 mm. The test focuses on investigation of crimping parameters and

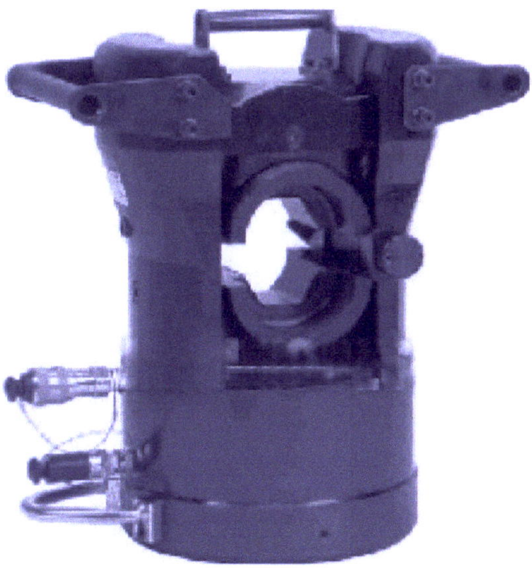

Figure 4.6 Conductor crimping machine.

crimping effect of carbon fiber composite core, therefore no crimping of external aluminum strand happens.

1. Preparation for crimping fittings
   a. Rinse the fittings with gasoline and wipe the interior of the pipe with clean cotton cloth.
   b. Carbon fiber reinforced composite core bar is evenly polished with abrasive paper, and the core bar surface is evenly polished to generate certain roughness from brightness, and the surface of the polished core bar is wiped out with clean cotton cloth.
   c. Insert core bar into the fitting for crimping.
2. Crimping and tension test of normal fittings

   Firstly, directly apply normal strain clamp used for ACSR with the same outer diameter to carbon fiber reinforced composite core and preliminarily explore feasibility of crimping connecting fittings applied to carbon fiber reinforced composite core.
   a. For the first group of samples, hydraulic machine and crimping mold are used to form three molds by pressing from the cable clamp steel anchor side to the conductor side. Pressure of the first mold and the second mold is 80 MPa and the pressure of the third mold is 40 MPa. Refer to Fig. 4.7 for strain clamp after crimping. The tensile force test is conducted after completion of crimping, the tensile force is 4.1 kN, and fracture pattern is breakage of the core bar in the fitting cavity.
   b. For the second group of samples, hydraulic machine and crimping mold are used to form three molds by pressing from the cable clamp steel anchor side to the conductor side; pressure of the first mold and the second mold is 60 MPa and the pressure of the third mold is 30 MPa. The tensile force test is conducted after

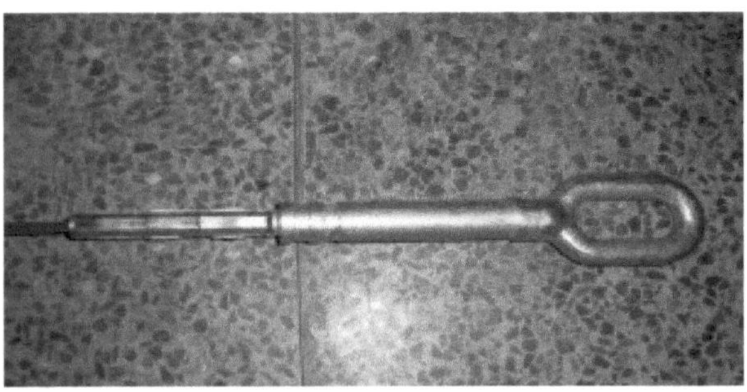

Figure 4.7 Strain clamp after crimping.

completion of crimping, the tensile force is 8.1 kN, and fracture pattern is breakage of the core bar in the fitting cavity.

c. For the third group of samples, hydraulic machine and crimping mold are used to form two molds by pressing from the cable clamp steel anchor side to the conductor side; pressure of the first mold and the second mold is 60 MPa and the pressure of the third mold is 40 MPa. The tensile force test is conducted after completion of crimping, the tensile force is 8.05 kN, and fracture pattern is breakage of the core bar in the fitting cavity.

d. Hydraulic machine and crimping mold are used for the fourth group; two molds from the cable clamp steel anchor side to conductor side are formed. The pressure of the first mold is 30 MPa and the pressure of the second mold is 20 MPa. Refer to Fig. 4.8 for the status of fittings after completion of crimping. The tensile force test is conducted after completion of crimping, then the tensile force is 46.75 kN and fracture pattern is core bar pulled out from the fitting cavity, shown in Fig. 4.9.

e. For the fifth group of samples, hydraulic machine and crimping mold are used to form two molds by pressing from the cable clamp steel anchor side to the conductor side. Pressure of the first mold and the second mold is 25 MPa and the pressure of the third mold is 20 MPa. The tensile force test is conducted after completion of crimping, then the tensile force is 45.9 kN and fracture pattern is core bar pulled out from the fitting cavity.

f. For the sixth group of samples, hydraulic machine and crimping mold are used to form two molds by pressing from the cable clamp steel anchor side to the conductor side. Pressure of the first mold

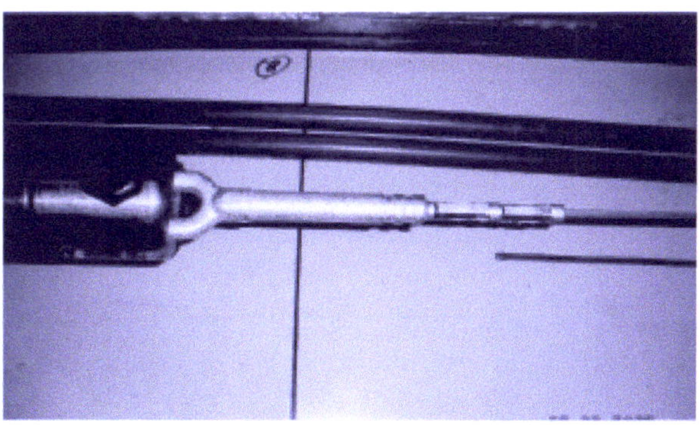

Figure 4.8 Status of fittings after completion of crimping.

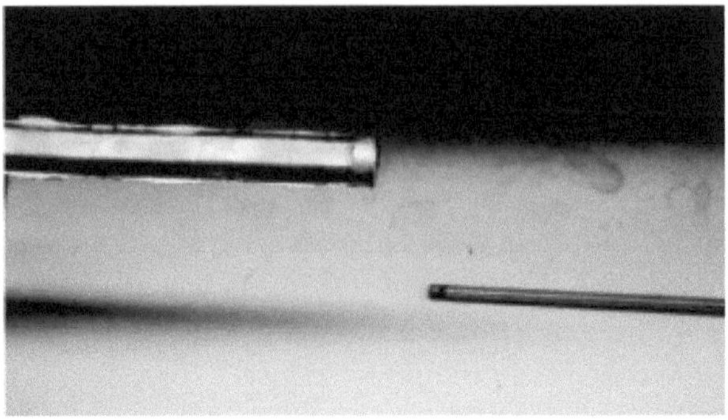

Figure 4.9 Status of core bar pulled out from fitting cavity.

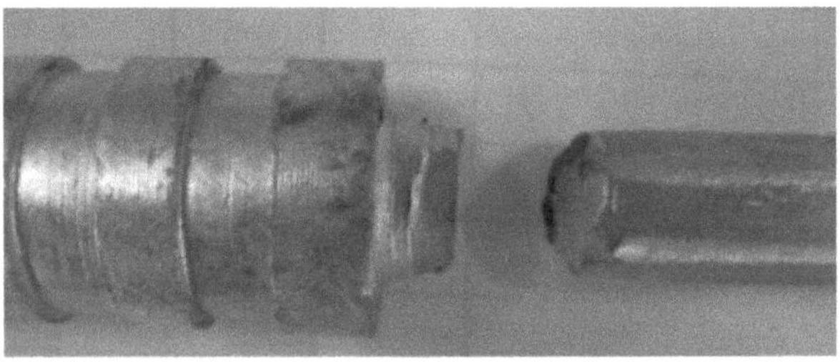

Figure 4.10 Tensile failure status of fittings.

and the second mold is 28 MPa and the pressure of the third mold is 20 MPa. The tensile force test is conducted after completion of crimping, then the tensile force is 46.25 kN and fracture pattern is core bar pulled out from the fitting.

g. For the seventh group of samples, hydraulic machine and crimping mold are used to form three molds by pressing from the cable clamp steel anchor side to the conductor side. Pressure of the first mold and the second mold is 28 MPa and the pressure of the third mold is 20 MPa. The tensile force test is conducted after completion of crimping, then the tensile force is 75.95 kN and fracture pattern is breakage of fittings, shown in Fig. 4.10.

h. For the eighth group of samples, hydraulic machine and crimping mold are used to form three molds by pressing from the cable clamp steel anchor side to the conductor side. Pressure of the first mold and the second mold is 28 MPa and the pressure of the third mold is 20 MPa. The tensile force test is conducted after completion of crimping, the tensile force is 78.05 kN, and fracture pattern is breakage of fittings.

i. For the ninth group of samples, hydraulic machine and crimping mold are used to form three molds by pressing from the cable clamp steel anchor side to the conductor side. Pressure of the first mold and the second mold is 28 MPa and the pressure of the third mold is 20 MPa. The tensile force test is conducted after completion of crimping, the tensile force is 76.15 kN, and fracture pattern is breakage of the fittings.

j. For the 10th group of samples, hydraulic machine and crimping mold are used to form three molds by pressing from the cable clamp steel anchor side to the conductor side. Pressure of the first mold and the second mold is 28 MPa and the pressure of the third mold is 20 MPa. The tensile force test is conducted after completion of crimping, the tensile force is 76.05 kN, and fracture pattern is breakage of the fittings.

It is shown that, on the basis of the test groups (a), (b), and (c), if crimping for carbon fiber reinforced composite core is conducted based on pressure value of normal conductor 60−80 MPa, the composite core will be crushed and the tensile force value will lower to about 8 kN. It is shown that, on the basis of the test groups (d), (e), and (f), for the abovementioned carbon fiber reinforced composite core and corresponding fittings, the pressure value of a single mold that is from 25 M to 30 MPa can give rise to the best crimping effect, and the tensile force value after crimping of two molds can reach to about 45 kN. It is shown based on the test groups (g), (h), (i), and (j), for the above carbon fiber reinforced composite core and corresponding fittings, pursuant to a pressure value (25 M−30 MPa) of a single mold, two molds replaced by three molds can further increase tension value, which can reach to about 75 kN.

On account of dimensional limits of normal strain clamps used for ACSR, crimping mold number cannot be further increased, thus project designs and processes should specially match crimping fittings for carbon fiber reinforced composite core to verify the feasibility of further increasing crimping mold number and tension value.

3. Crimping and tensile force test of special fittings

For the abovementioned carbon fiber reinforced composite core, special matched crimping fittings for lengthening crimping part are designed, and the crimping test of carbon fiber reinforced composite core is further conducted. The test results are as follows, namely:

a. For the first group of samples, hydraulic machine and crimping mold are used to form four molds by pressing from the cable clamp steel anchor side to the conductor side, and pressure of the first mold, the second mold, and the third mold is 40 MPa and the pressure of the fourth mold is 20 MPa. The tensile force test is conducted after completion of crimping, and the tensile force is 101 kN and fracture pattern is breakage of the core bar in the fitting cavity.

b. For the second group of samples, hydraulic machine and crimping mold are used to form four molds by pressing from the cable clamp steel anchor side to the conductor side, and pressure of the first mold, the second mold, and the third mold is 40 MPa and the pressure of the fourth mold is 20 MPa. The tensile force test is conducted after completion of crimping, then the tensile force is 101 kN and the fracture pattern is breakage of the core bar at the exit of the fitting cavity, shown in Fig. 4.11.

c. For the third group of samples, hydraulic machine and crimping mold are used to form four molds by pressing from the cable clamp steel anchor side to the conductor side, and pressure of the first mold, the second mold, and the third mold is 34 MPa and the pressure of the fourth mold is 20 MPa. The tensile force test is conducted after completion of crimping, and the tensile force is 112.05 kN and fracture pattern is breakage of the core bar at the exit of the fitting cavity.

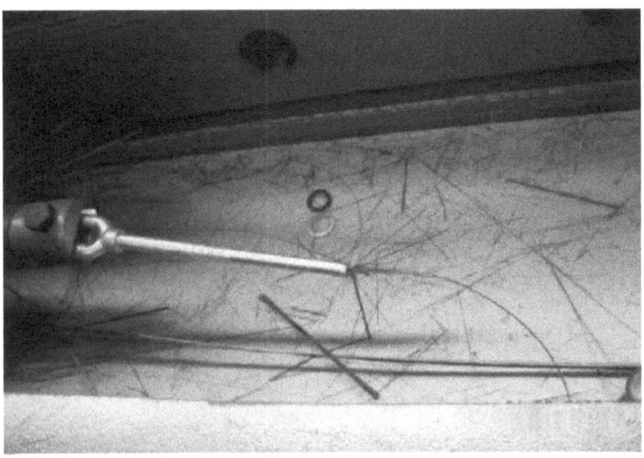

Figure 4.11 Status of core bar breakage at the exit of fitting cavity.

d. For the fourth group of samples, hydraulic machine and crimping mold are used to form four molds by pressing from the cable clamp steel anchor side to the conductor side, and pressure of the first mold, the second mold, and the third mold is 30 MPa and the pressure of the fourth mold is 20 MPa. The tensile force test is conducted after completion of crimping, and the tensile force is 105 kN and fracture pattern is breakage of the core bar at the exit of the fitting cavity.

e. For the fifth group of samples, hydraulic machine and crimping mold are used to form four molds by pressing from the cable clamp steel anchor side to the conductor side, and pressure of the first mold, the second mold, and the third mold is 28 MPa and the pressure of the fourth mold is 20 MPa. The tensile force test is conducted after completion of crimping, and the tensile force is 111.25 kN and fracture pattern is breakage of the core bar at the exit of the fitting cavity.

f. For the sixth group of samples, hydraulic machine and crimping mold are used to form four molds by pressing from the cable clamp steel anchor side to the conductor side, and pressure of the first mold, the second mold, and the third mold is 24 MPa and the pressure of the fourth mold is 20 MPa. The tensile force test is conducted after completion of crimping, and the tensile force is 111.25 kN and fracture pattern is breakage of the core bar at the exit of the fitting cavity. The fitting is cut apart after the sample in group 6 is broken, and the internal form is shown in Fig. 4.12.

g. For the seventh group of samples, hydraulic machine and crimping mold are used to form four molds by pressing from the cable clamp steel anchor side to the conductor side, and pressure of the first mold, the second mold, and the third mold is 16 MPa and the pressure of the fourth mold is 16 MPa. The tensile force test is conducted after completion of crimping, then the tensile force is 89.9 kN, and fracture pattern is core bar pulled out from fitting cavity.

h. For the eighth group of samples, hydraulic machine and crimping mold are used to form four molds by pressing from the cable clamp steel anchor side to the conductor side, and pressure of the first mold, the second mold, and the third mold is 20 MPa and the pressure of the fourth mold is 20 MPa. The tensile force test is conducted after completion of crimping, then the tensile force is

Figure 4.12 After incision of fittings.

87.7 kN, and fracture pattern is core bar pulled out from fitting cavity.

i. For the ninth group of samples, hydraulic machine and crimping mold are used to form four molds by pressing from the cable clamp steel anchor side to the conductor side, and pressure of the first mold, the second mold, and the third mold is 20 MPa and the pressure of the fourth mold is 20 MPa. The tensile force test is conducted after completion of crimping, then the tensile force is 105.35 kN, and fracture pattern is core bar pulled out from the fitting cavity.

j. For the 10th group of samples, hydraulic machine and crimping mold are used to form four molds by pressing from the cable clamp steel anchor side to the conductor side, and pressure of the first mold, the second mold, and the third mold is 22 MPa and the pressure of the fourth mold is 20 MPa. The tensile force test is conducted after completion of crimping, then the tensile force is 105.35 kN, and fracture pattern is core bar pulled out from the fitting cavity, and the outer resin is damaged.

k. For the 11th group of samples, hydraulic machine and crimping mold are used to form four molds by pressing from the cable clamp steel anchor side to the conductor side, and pressure of the first mold, the second mold, and the third mold is 24 MPa and the pressure of the fourth mold is 20 MPa. The tensile force test is

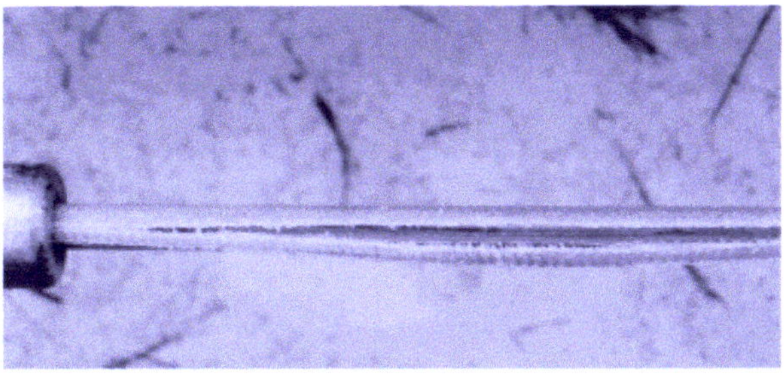

Figure 4.13 Status at the time of tensile force reaching to 109 kN.

conducted after completion of crimping, and the tensile force is 113.95 kN and fracture pattern is the core bar pulled out from the fitting cavity with outer resin damage. Refer to Fig. 4.13 for the status before breakage.

1. For the 12th group of samples, hydraulic machine and crimping mold are used to form four molds by pressing from the cable clamp steel anchor side to the conductor side, and pressure of the first mold, the second mold, and the third mold is 26 MPa and the pressure of the fourth mold is 20 MPa. The tensile force test is conducted after completion of crimping, then the tensile force is 113.3 kN and fracture pattern is core bar pulled out from the fitting cavity, and the outer resin is damaged.

The tests indicated that forming four molds by pressing from the cable clamp steel anchor side to the conductor side can further increase tensile force value, which reaches to above 105 kN and increases by about 40% compared with that of three molds. However, subsequent tests show that the tensile force value is difficult to increase further even if mold crimping number is increased.

4. Dissection of fittings for inspection after the tensile force test

To further confirm faulted condition of carbon fiber reinforced composite core after crimping with different pressure and tensile force test, dissect and inspect various samples. For status comparison after fitting crimping (including samples before and after tensile force test), refer to Fig. 4.14.

After crimping of fittings under different pressures, status of direct dissection (namely status without tensile force test) and status of dissection after the tensile force test are shown in Figs. 4.15–4.19, respectively.

Dissection results show that at the time of crimping of fittings, if the pressure of a single mold is less than 24 MPa, the core bar is easily pulled

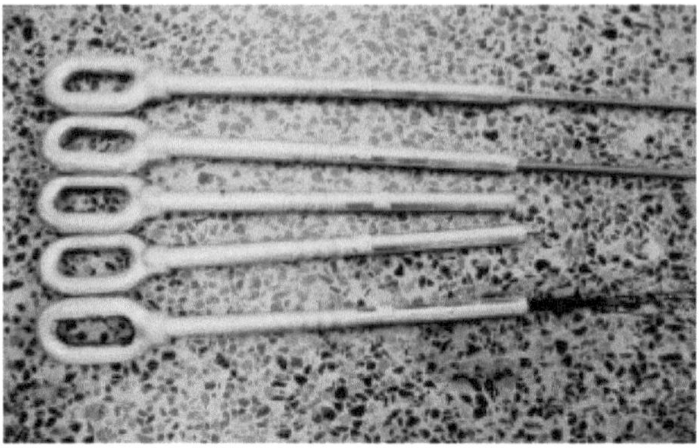

Figure 4.14 Status comparison after crimping of fittings.

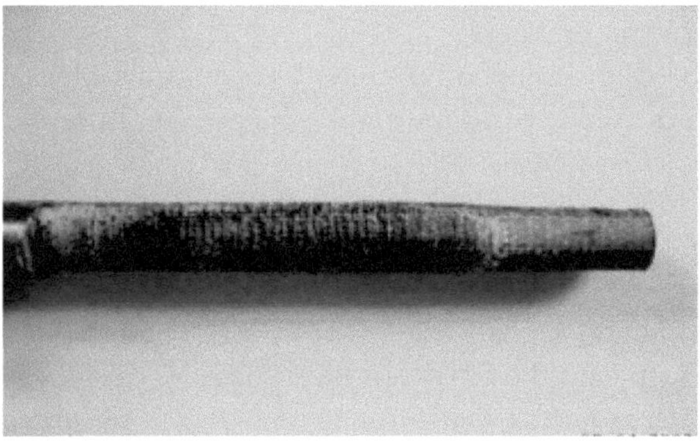

Figure 4.15 Status of core bar without tensile force exerted under the pressure of 24 MPa.

Figure 4.16 Status of core bar with tensile force of 113.95 kN under the pressure of 24 MPa.

Figure 4.17 Status of core bar without tensile force exerted under the pressure of 22 MPa.

Figure 4.18 Status of core bar with tensile force of 114.6 kN under the pressure of 22 MPa.

Figure 4.19 Status of core bar with tensile force of 113.3 kN under the pressure of 26 MPa.

out from the fitting cavity in the process of the tensile force test; if the pressure is more than 30 MPa, the composite core is easily crushed with significant reduction of tensile force of the core bar and connecting fittings.

To sum up, crimping construction parameters regarding carbon fiber reinforced composite core and matched crimping connecting fittings based on independent research and development in China are as follows. The pressure scope of a single mold at the time of crimping is 25 M−30 MPa; the median 28 MPa is advisable at the time of actual operation, but the pressure of the final mold in the conductor side (i.e., exit side of fitting cavity) is 20 MPa. For carbon fiber reinforced composite core with diameter of 7.5 mm, it is necessary to form four molds by pressing from the cable clamp steel anchor side to the conductor side.

4.3.1.2 Other Tests

1. Grip strength test at room temperature

   Grip strength at room temperature of the conductor (including aluminum strand) and crimping fittings is 169 kN, which is 41% higher than calculated tensile force.

2. Grip strength test at room temperature

   Test process: Initial tension is 5 kN, which is kept for 20 min. Then heat the temperature to 120°C, and increase conductor tension to 60 kN when the temperature reaches to 120°C. Carry out the tensile force test for the conductor for 4 h.

   The grip strength of conductor and crimping fittings at 120°C is 152 kN, which is 27% higher than calculated tensile force and 10% lower than grip strength at room temperature. The grip strength of conductor and crimping fittings at 150°C is 149 kN, which is 24% higher than calculated tensile force and 12% lower than grip strength at room temperature. The grip strength of conductor and crimping fittings at 170°C is 149.6 kN, which is 25% higher than calculated tensile force and 11% lower than grip strength at room temperature (refer to Fig. 4.20 for test process). The above data is similar to the test result of wedge joint fittings.

3. Fatigue vibration test

   To inspect the capacity of crimping straight connecting pipe and strain clamp bearing alternating stress, install straight connecting pipe in the middle of the conductor and put the connecting pipe close to vibration excitation device in the process of fatigue vibration test (refer to Fig. 4.21). The test results indicated that both carbon fiber composite core

Figure 4.20 Tensile force test with connecting fittings at 170°C.

Figure 4.21 Fatigue vibration test with connecting fittings.

aluminum conductor and crimping connecting pipe can smoothly pass the test of dynamic fatigue vibration for 30 million times. And after the sample passes the fatigue vibration test, no decline in grip strength is seen.

4. Test of passing the tackle

To grasp related parameters regarding crimping connecting pipe passing the tackle in the process of continuous tension stringing of carbon fiber composite core aluminum conductor, carry out a tackle test against a composite sample of carbon fiber composite core aluminum conductor JLNR/F2B-300/65 (diameter of composite core: 9 mm) and crimping connecting pipe and test core shrinking of the conductor.

The following parameters are confirmed based on test results. In case of long-distance continuous tension stringing based on combination of carbon fiber composite core aluminum conductor JLNR/F2B-300/65 with diameter of composite core being 9 mm, and matched connecting pipe, it is necessary to select a tackle with diameter of not less than 822 mm, the envelope angle (approximately equal to line angle) of the conductor passing the tackle should not be more than 25 degrees, and the connecting pipe (covered by steel armor for protection) should not pass the tackle more than 20 times. As for construction machinery of carbon fiber composite core aluminum conductor with composite core diameter of less than 9 mm and the construction machine of matched connecting pipe, please select by referring to the parameters above.

Due to condition restrictions, the size of steel armor for protection selected for the test is too large, that is to say that the diameter of steel armor for protection is slightly larger than the groove width of 822-mm tackle; as a result, it is easy for the steel armor to produce a large impulsive force when it passes the tackle. The larger the envelope angle of the conductor passing the tackle, the larger the impulsive force produced will be, and it is easy to incur breakage of the composite core. If the size of the steel armor can well match with tackle size, the above parameters can be properly relaxed.

5. Temperature rise test

Using metal monofilament, pretwisted patch rods, repair sleeve, pretwisted connecting strip, wedge joint straight connecting pipe, crimping straight connecting pipe, and other fittings to repair the conductor, mechanical performance and behavior of electricity (namely through-current capability) of the conductor shall be recovered at the same time. Unreasonable design of straight connecting pipe, product quality problems, improper repairing technology of the conductor, etc. will lead to conductor repair quality problems, including large contact resistance of connections, which causes severe heat in the straight connecting pipe during the operational process.

The temperature rise test when large currents are going through carbon fiber composite core aluminum conductor repaired by crimping straight connecting pipe can provide a basis for design, construction, operation, and maintenance of carbon fiber composite core aluminum conductor.

Table 4.1 Temperature rise test data regarding conductor and fittings

Applied current (A)	Test point	Monitoring temperature (°C)
605	Conductor	70
	Strain clamp	66
	Straight connecting pipe	67
842	Conductor	100
	Strain clamp	88
	Straight connecting pipe	87

a. Test methods

   i. Select JLNR/F2B-240/45 carbon fiber composite core aluminum conductor with length of 20 m as sample with crimping strain clamp installed on both ends.

   ii. Cut off the conductor in the place 5 m away from one end of the conductor, and use crimping straight connecting pipe to repair the part cut.

   iii. Embed thermocouple probe into conductor, connecting pipe, strain clamp, etc.

   iv. Connect strain clamps on both ends of the conductor to output cable of current rise device and monitor temperatures of conductor, connecting pipe, and strain clamp by thermal couple.

b. Test results. Refer to Table 4.1 for test temperature of straight connecting pipe, strain clamp, and conductor. The temperature of the straight connecting pipe and strain clamp is lower than that of the conductor because the flow-through section and heat radiating area of crimping strain clamp and straight connecting pipe are larger than that of the conductor. Such test data conforms to design requirements of fittings. The test indicates that both crimping strain clamp and crimping straight connecting pipe can effectively recover electric conductivity and through-current capability of carbon fiber composite core aluminum conductor.

4.3.2 Case Regarding Continuous Tension Stringing of Carbon Fiber Composite Core Aluminum Conductor

After the research and development of crimping connecting fittings of carbon fiber composite core aluminum conductor in China by SECRI, long-distance continuous tension stringing was realized for the first on the

Figure 4.22 Straight connecting pipe of carbon fiber composite core aluminum conductor with steel armor for protection.

Figure 4.23 Carbon fiber composite core aluminum conductor with straight connecting pipe passes the tackle.

500 kV line from Zhangnan to Mentougou and from Mentougou to Fangshan, which is in northern China.

JLNR/F2B-240/45 conductor is used and is connected to several matched straight connecting pipes (refer to Fig. 4.22), forming a testing line. Two tension stringing sections are mountain areas with complicated terrain, and the line has many corners with the maximum corner (approximately equal to envelope angle of conductor passing the tackle) about 50 degrees. When the corner is more than 25 degrees, double tackle is adopted (refer to Fig. 4.23). The straight connecting pipe passes the tackle for nine times and six times, respectively. At the time of paying-off, the conductor tension on the side of the tension field is 10–20 kN.

Figure 4.24 Recycling of samples.

Recycling of samples after the test of passing the tackle on the scene is shown in Fig. 4.24. For partial samples, remove protective steel armor outside of the straight connecting pipe and strip off conductor aluminum strand for inspection, and no damage happens to the composite core. For partial samples, conduct tensile force test, and no decline of grip strength happens to the conductor and straight connecting pipe.

4.4 PRETWISTED WEDGE JOINT COMBINED TYPE CONNECTING FITTINGS

4.4.1 Research and Development of Pretwisted Wedge Joint Combined Type Connecting Fittings

In recent years, for normal ACSR, operation and maintenance more and more widely uses preformed armor rods with full recovery of tension as connecting fittings for repairing the conductor during the period of operation. Internal steel core and external aluminum strand are connected by aluminum alloy preformed armor rods, thus not too many, too large, or too heavy auxiliary means are needed in the installation process, and such method is welcomed by maintenance workers. For carbon fiber composite core aluminum conductor, operation and maintenance units propose research and development of similar user-friendly pretwisted connecting fittings. At present, preformed armor rod connecting fittings are only applicable to straight connecting fittings.

1. Solution I: Preformed armor rods are used for connection internally and externally. The scheme shares the same connection format with pretwisted connecting fittings of ACSR, that is to say that preformed

armor rods are used for twining and connecting carbon fiber reinforced composite core internally and preformed armor rods are used for twining and connecting aluminum strand externally. However, at the time of the test, it is found that carbon fiber reinforced composite core is a single rod with a relatively smooth surface, large friction is hard to form between aluminum alloy preformed armor rods and the optical wand, preformed armor rods are hard to twine with the composite core just like twining with steel strand, and the tensile force after repair cannot be ensured; as a result, the scheme is abandoned.

2. Solution II: Pretwisted wedge joint combined type connection means wedge joint fittings are internally used to connect carbon fiber reinforced composite core and preformed armor rods are externally used to twine and connect the aluminum strand; however, the outer diameter of the wedge joint fitting connecting carbon fiber reinforced composite core is always larger than the outer diameter of the aluminum strand of the conductor (refer to Fig. 4.25), so preformed armor rods externally twining aluminum strand cannot be continuously twined with wedge joint fittings at the section of fitting, but such preformed armor rods can only go through the outer side of wedge joint fittings to another side of the wedge joint fittings, and then can continuously twine with aluminum strand (refer to Fig. 4.26). The above connection type can reach to tensile strength and through-current

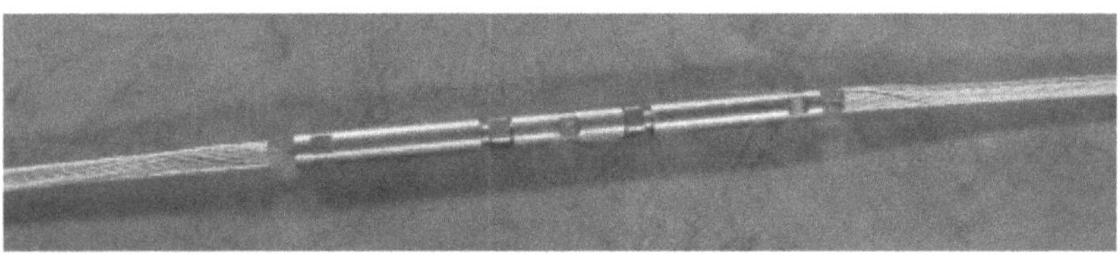

Figure 4.25 Wedge joint fittings of connecting carbon fiber reinforced composite core.

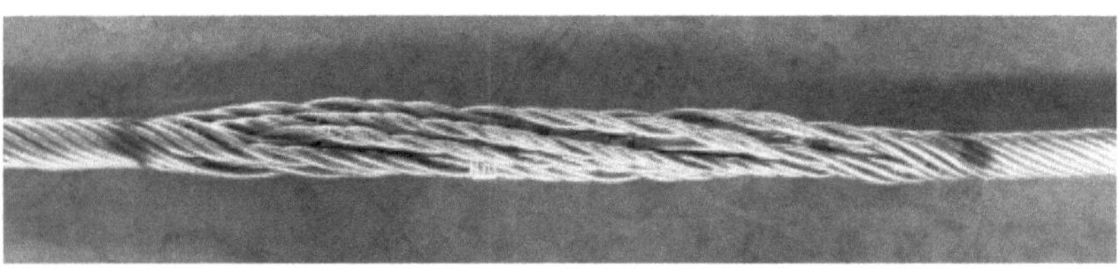

Figure 4.26 Going through preformed armor rods along the outer side of wedge joint fittings.

capability of the original conductor, but fails in appearance, and the protection effect of preformed armor rod carbon fiber reinforced composite core is doubtful, thus the scheme is abandoned.

3. Solution III: This scheme involves pretwisted wedge joint combined type connections, that is to say that wedge joint fittings are used to connect carbon fiber reinforced composite core internally, and preformed armor rods are used to twine aluminum strand externally. One pair of filling preformed armor rods is added due to the problem of the outer diameter of wedge joint fittings connecting carbon fiber reinforced composite core always being larger than that of aluminum strand; these are twisted on the aluminum strand on both sides of the wedge joint fittings to make the outer diameter of twisted aluminum strand equal to or slightly larger than that of the wedge joint fittings (refer to Figs. 4.27 and 4.28). Finally, connecting preformed armor rods are directly twisted on wedge joint fittings and filling preformed armor rods on both sides (refer to Figs. 4.29 and 4.30). The

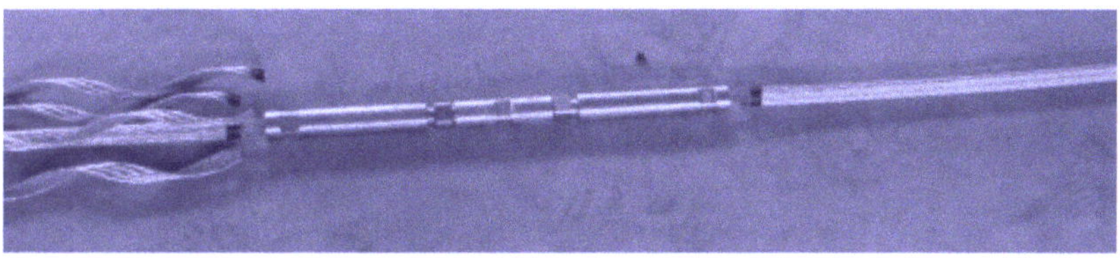

Figure 4.27 Filling preformed armor rods to be installed.

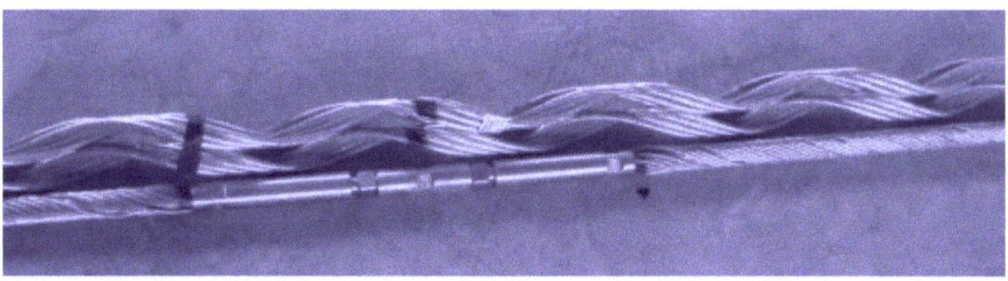

Figure 4.28 Filling preformed armor rods after installation and connecting preformed armor rods to be installed.

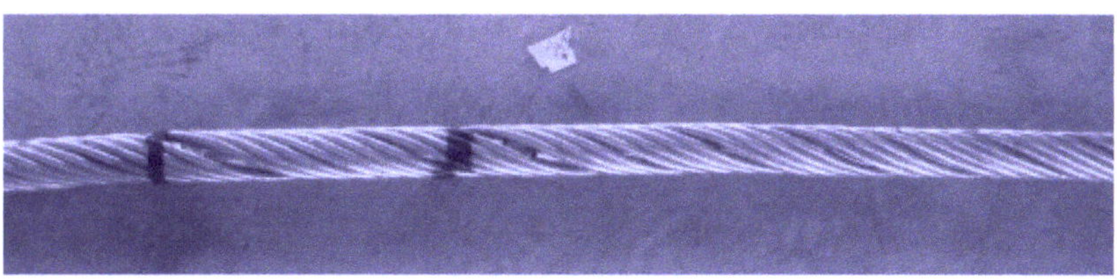

Figure 4.29 Continuous preformed armor rods after installation.

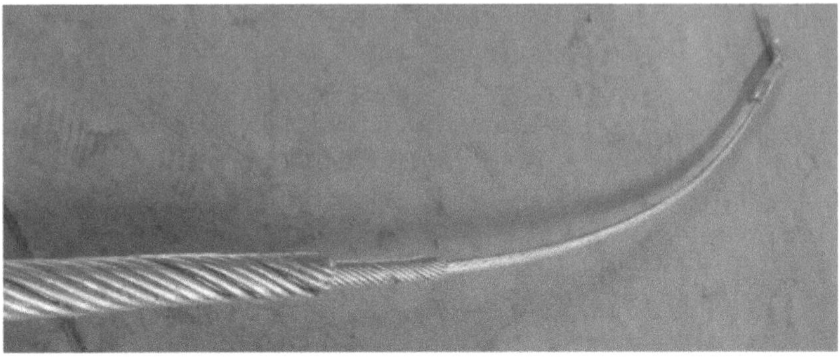

Figure 4.30 Carbon fiber composite core aluminum conductor connected by pretwisted wedge joint combined type connecting fittings.

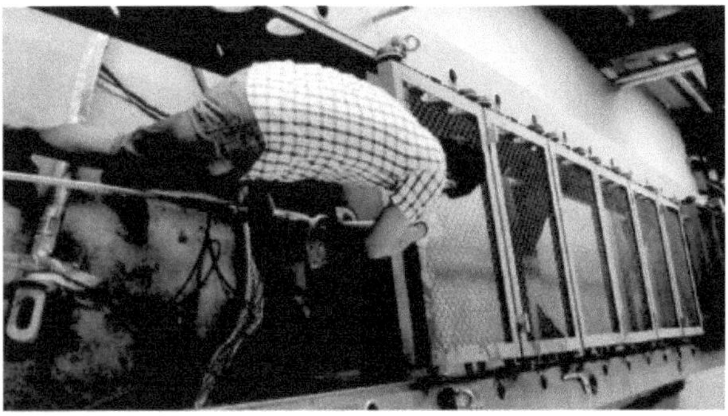

Figure 4.31 Tensile force test.

connection part formed based on the scheme is specious, has small dimension, and provides protection for carbon fiber reinforced composite core obviously superior to that of Scheme II; thus, finally this scheme is confirmed for use.

4.4.2 Conductor and Fittings Tensile Force Test

For the pretwisted wedge joint combined type connecting fittings confirmed by the above scheme, the continuous strength required shall be no lower than 95% of that of the original conductor. The specification of heat-resistant aluminum alloy twisted conductor of carbon fiber reinforced composite core use for tensile force test is JLNR/F2B-240/45 (sectional area of heat-resistant aluminum alloy is 242.8 mm^2; diameter of carbon fiber composite core is 7.5 mm with sectional area of 44.2 mm^2), rated tensile force is 125.5 kN, and 95% of rated tensile force is 120 kN. The tensile force test is shown in Fig. 4.31.

For the sample of heat-resistant aluminum alloy twisted conductor of the above carbon fiber reinforced composite core connected by pretwisted wedge joint combined type connecting fittings, the tensile and tolerance tests shall be conducted in light of the following procedures:

1. Within 45–90 s, increase sample tension to 60 kN from 0 based on constant speed, which is kept for 60 s.
2. Increase sample tension to 120 kN from 60 kN based on constant speed, which is kept for 60 s. No cracking or other damage happens to three samples after the test, and these three samples pass the test.

4.4.3 Temperature Rise Test of Conductor and Fittings

For the sample of heat-resistant aluminum alloy twisted conductor of the above carbon fiber reinforced composite core connected by pretwisted wedge joint combined type connecting fittings, its through-current capability is required to be not lower than that of the original conductor; the specific assessment method is that when certain current passes the sample and the temperature becomes stable, the temperature rise of heat-resistant aluminum alloy twisted conductor for connection shall not be more than that of original conductor heat-resistant aluminum alloy. Temperature rise test is shown in Fig. 4.32.

In the process of the actual test, the sample is passed by 654 A power current. After sample temperature becomes stable and when heat-resistant aluminum alloy of original conductor reaches to 66°C, the temperature of heat-resistant aluminum alloy preformed armor rods for connection is

Figure 4.32 Temperature rise test.

only 36°C, which is less than that of the heat-resistant aluminum alloy of the original conductor.

In addition, eolian vibration fatigue test for 30 million times is conducted for the sample of the above carbon fiber reinforced composite core heat-resistant aluminum alloy twisted conductor connected by pre-twisted wedge joint combined type connecting fittings; as a result, no damages such as broken strands are seen in the connections, and product design requirements are met.

4.5 COMBINED TACKLE FOR PAYING-OFF AND CONSTRUCTION UNDER SPECIAL CONDITIONS

4.5.1 Study Target

Although carbon fiber composite core aluminum conductor has obvious technical and economic advantages as theoretically applied to a long-span line, no practical application cases using it have arisen at home or abroad. Firstly, certain difficulty arises in the research and development of super strength carbon fiber composite core aluminum conductor applicable to long-span lines, and magnesium, aluminum, and silicon alloy is hard to well match with carbon fiber reinforced composite core in terms of strength. Secondly, since carbon fiber reinforced composite core is of organic composite material, its softness is relatively low compared with that of the normal steel core. For carbon fiber composite core aluminum conductor with super strength, the risk of large envelope angle passing the tackle based on large tension in the process of tension stringing will greatly increase and normal construction equipment and construction parameters will no longer apply to long-span lines. Therefore, it is a must to research and develop specially matched construction equipment (such as special tackles) based on the specific performance of carbon fiber composite core aluminum conductor, and form a set of construction technologies for safe paying-off under bad conditions such as long span and large height difference to lay the foundation of construction and application of carbon fiber composite core aluminum conductor under special conditions.

4.5.2 Design, Research, and Development of Combined Tackle for Paying-Off

4.5.2.1 Technical Features of Ordinary Tackle for Paying-Off

In general, tackle for paying-off used for overhead lines has large diameter; its breast wheel can be cast steel wheel or nylon wheel; its basic usage

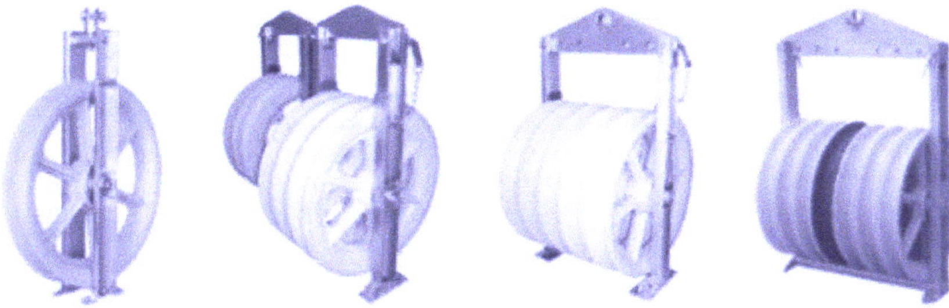

Figure 4.33 Tackle for paying-off.

Figure 4.34 Double tackle for paying-off.

is leading and placing single conductor, double divided conductor, and quadripartion conductor; and it is applicable to tension stringing, as shown in Fig. 4.33.

In general, tackle for paying-off has wheel bottom diameter of 400–800 mm and wheel width of 75–125 mm. Tackles with different sizes are selected according to different conductor models. The bigger the sectional area of the conductor is, the bigger the diameter, the bigger the radius needed for paying-off, the bigger the diameter of the selected tackle for paying-off, and the bigger the wheel width will be.

At the time of tension stringing, if the tower corner or envelope angle of the conductor passing the tackle is big (more than 30 degrees), it is necessary to use a double tackle for paying-off (refer to Fig. 4.34) to ensure no conductor is damaged in the process of tension stringing.

4.5.2.2 The Requirements of Carbon Fiber Composite Core Aluminum Conductor Tension Stringing Under General Conditions

As stated in Section 3.1, under general conditions, tackle requirements for carbon fiber composite core aluminum conductor tension stringing are as follows:

1. When the envelope angle of the conductor passing single tackle for paying-off is less than 30 degrees and quantity of the tower continuously passed fails to exceed 20 (conforming to the requirements for tension stringing of ordinary conductor), conductor tension is generally less than 30 kN. For the conductor using the composite core of 9 mm, it is required to use 800 mm and above tackle (such as tackle with trench bottom diameter of 822 mm). With the decrease in diameter of composite core, the tackle size will decrease accordingly.
2. When the envelope angle of the conductor passing the tackle for paying-off is more than 30 degrees, but less than 60 degrees, it is required to use double tackle for paying-off and that quantity of tower continuously passed fails to exceed 20. In general, conductor tension is less than 30 kN. For the conductor using the composite core of 9 mm, it is required to use 800 mm and above tackle (such as tackle with trench bottom diameter of 822 mm). With the decrease in diameter of composite core, the tackle size will decrease accordingly.

However, in the process of actual line construction, the envelope angle of the conductor passing the tackle for paying-off is always more than 60 degrees, and the conductor tension is often more than 30 kN, including some lines with a big height difference, large corner, and long span. The following situations are encountered by high voltage side voltage controller in the actual engineering of carbon fiber composite core aluminum conductor: for No. 89 tower (220 kV) of Hongman Line in Urumchi, the envelope angle reaches to 70 degrees; for No. 41–54 towers (220 kV) of Taoping Line, the envelope angle reaches to 95 degrees; for capacity expansion modification works of Meishan 110 kV padpower double double-circuit line in Nanjing, capacity-increase modification works of 35 kV line from Mucun, Liulin County, Lvliang City, Shanxi to Dengjiazhuang, Liulin County, Lvliang City, Shanxi, reverse traction is needed for the conductor due to limitation of tension field, that is to say that the envelope angle of the conductor passing the tackle reaches to 180 degrees. Ordinary single tackle and double tackle cannot meet the requirement of continuous tension stringing of carbon fiber composite core aluminum conductor under such special conditions, and it is a must to research and develop special construction equipment so as to ensure no damage to carbon fiber composite core aluminum conductor in the process of tension stringing.

4.5.2.3 Research and Development of Combined Tackle for Paying-Off

The overall design of combined tackle for paying-off for carbon fiber composite core aluminum conductor is as follows. In accordance with static winding diameter allowed by carbon fiber composite core aluminum conductor and possible damage to composite core caused by lateral extrusion force exerted by the tackle to the conductor at the time of the conductor with tension passing the tackle and considering a certain safety margin, the design can meet tackle for paying-off of carbon fiber composite core aluminum conductor under the conditions of large envelope angle (60–180 degrees) and large laying tension.

The bigger the diameter of carbon fiber reinforced composite core is, the bigger the winding diameter allowed will be, so the bigger the tackle for paying-off needed will be. At present, carbon fiber reinforced composite core with maximum diameter of the investment bank is 9 mm and the maximum specification of carbon fiber reinforced composite core listed in GB/T 29324.2012 *Fiber-reinforced Resin Composite Core for Overhead Conductor* and Q/GDW 1851 *Carbon Fiber Composite Core Overhead Conductor* is 11 mm, thus special tackle for paying-off shall be designed pursuant to the conductor made by twining carbon fiber reinforced composite core with diameter of 11 mm. In case of being able to meet the requirement of its tension stringing, the requirement of carbon fiber composite core aluminum conductor with diameter of composite core being less than 11 mm can be met, too.

The maximum twining diameter allowed by carbon fiber reinforced composite core with diameter of 11 mm is 55 times of that of its core bar, namely $11 \times 55 = 605$ (mm), however, such value is static winding data without such influence factors as extrusion, laying tension, and speed of paying-off of external aluminum strand, as well as times of the conductor passing the tackle, thus design dimensions of the tackle shall be greatly relaxed on the basis of the above static winding diameter. In accordance with long-span actual construction experience, select the maximum laying tension of 60 kN. The maximum envelope angle of the conductor passing the tackle shall be considered based on 180 degrees, which is the most severe. Based on the above factors, confirm that basic diameter of this special tackle is three times that static winding diameter, namely 1800 mm, to meet the tension stringing requirement of carbon fiber composite core aluminum conductor with core bar diameter of not more than 11 mm safely passing the tackle based on any envelope angle.

In the process of actual research and development, it is found that in case of simply making tackle in line with the diameter of 1800 mm, the following problems exist:

1. Great processing difficulty. In general, the diameter of ordinary tackle for paying-off does not exceed 900 mm. For the tackle with diameter of 1800 mm, it is a must to redesign the mold with long production phase and high price. In case of production based on complete machining, manufacturing difficulty is great and it is hard to realize volume production.
2. Great application difficulty. When the tackle diameter reaches to 1800 mm, based on current tackle parameters, it is deduced that its quality reaches to 700 kg and its dimension will greatly increase; as a result, transportation is inconvenient for construction in areas such as mountains and ridges. For tackle for paying-off possibly hung and fixed based on various statuses, use difficulty will increase.

To overcome the above difficulties, the technical demands for this special tackle for paying-off are concluded based on usage requirements as follows:

1. Contact and relative motion between conductor and tackle based on friction rolling happen.
2. Bending radius of the conductor passing the tackle is not less than 900 mm (diameter: 1800 mm).
3. Tackle quality shall be as light as possible and tackle size is relatively small for the sake of facilitating transportation and usage.

On the basis of the above technical demand, the realization scheme of the special tackle block is proposed as follows. On the circular section with radius of 900 mm (diameter: 1800 mm), fix several small pulleys with tangency between surface and arc line and form several discrete cambered surfaces. Small pulleys are spread over the same arc with radius of 900 mm, namely one big tackle with diameter of 1800 mm consisting of several small pulleys that can restrict bending radius of the conductor at the time of paying-off of the single big tackle with diameter of 1800 mm, as shown in Fig. 4.35. Considering that the maximum envelope angle is 180 degrees at the time of paying-off (namely paying-off of the conductor in negative direction), the tackle block constituted by small pulleys only needs to form a semicircle with radius of 900 mm, whose size decreases

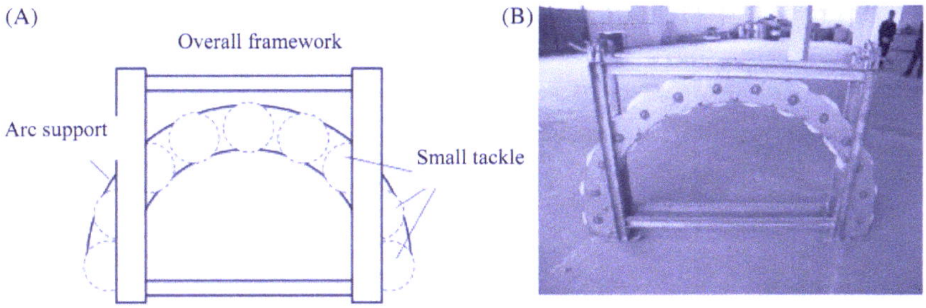

Figure 4.35 Combine tackle for paying-off with maximum envelope angle of 180 degrees.
(A) Schematic diagram of combined tackle for paying-off. (B) Material object of combined tackle for paying-off.

by 50% in comparison with the single complete big tackle with radius of 900 mm.

Arc support with radius of 900 mm is made of steel plate with certain thickness. Small tackle adopts nylon material similar to that of ordinary tackle for paying-off. Groove size of small tackle is consistent with that of single tackle for paying-off to ensure that the conductor, connecting fittings, and protective steel armor smoothly pass. All small pulleys are fixed to corresponding positions of arc support as shown in Fig. 4.35A. Outer frame of the tackle is made by welding channel steel and is assembled to the steel plate of the arc support, which ensures rigidity of the tackle as well as facilitates hanging of the tackle.

Conduct trial production of combined tackle for paying-off according to the above design scheme and gradually optimize design and processing schemes in the process of trial production.

1. At the time of initial trial production, nylon-6 manufacturing material for ordinary tackle for paying-off is selected for the small pulley, however, because the perimeter of the small tackle is far less than that of the single big tackle with radius of 900 mm, times of contact and friction between the surface of small tackle and the conductor within the unit time at the time of paying-off is 30–50 times of that of big pulley. Severe heating happens to surface material of the tackle groove and obvious conductor indentation and different levels of surface falling out arises at the time of conductor continuously and quickly passing the small tackle, which in actual application results in a difficulty in meeting demand in the process of paying-off. If manufacturing material of the small tackle is replaced with steel, on one hand, it is

estimated that the quality will increase by 80%, which goes against handling and usage; on the other hand, abrasion loss will certainly increase at the time of contact between the conductor and the pulley, that is to say that the conductor may be damaged. The scheme ultimately determined uses nylon-66 materials with good heat-resistant quality for the small pulley, and the test effect is good.

2. In the simulated paying-off test for the first trial production, the combined tackle encounters the problem of damage to conductor composite core. After the observation, it is found that imprinting on the surface of the small tackle and imprinting incurred by conductor extrusion are inconsistent and the direction of rotation of the tackle is inconsistent with conductor tendency. The reason is that machining precision of the small tackle is relatively low and positioning accuracy of the small tackle at the time of installation is relatively low, meaning that consistency of diameters of all the small pulleys cannot be ensured and installation location cannot be precisely controlled to be on the cambered surface with radius of 900 mm. In the process of subsequent trial production, strengthen accuracy control to solve the problem.

4.5.2.4 The Test of Passing the Tackle Against Combined Tackle for Paying-Off

1. Transformation of test system of passing the tackle

The test system of passing the tackle set forth in Section 3.1 of this book can simulate working conditions of tension stringing on the field of carbon fiber composite core aluminum conductor. By changing the suspension height of the tackle, the test system can simulate the scope of the envelope angle of the conductor passing the tackle at 15–60 degrees, which meets the test of passing the tackle under general line condition. However, the construction condition with envelope angle passing the tackle being over 60 degrees cannot be simulated.

For laboratory simulation of working conditions of conductor tension stringing incurred under the above special line construction condition, the transformation against the test system of passing the tackle in Section 3.1 of the book is as follows: Add one footing tackle on both sides of the test base frame to substantially increase the envelope angle scope of the system simulation test. Based on these two added footing pulleys, the envelope angle of the conductor passing the tackle can increase to 145 degrees from 60 degrees (shown in Fig. 4.36).

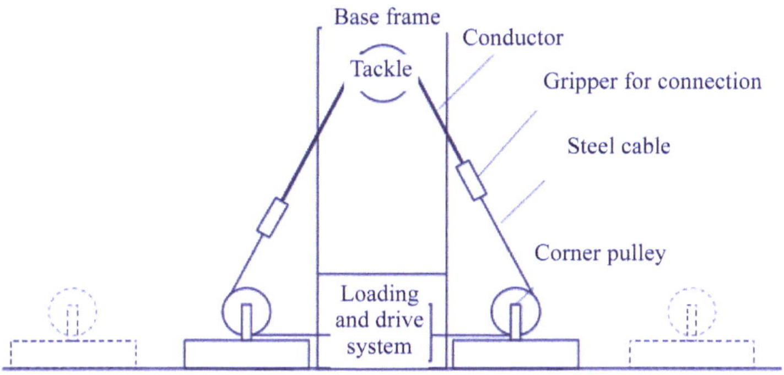

Figure 4.36 Experimental system regarding conductor passing the tackle with envelope angle increased.

2. Single-tackle test for paying-off

State grid corporation of china enterprise standard-Q/GDW 1851 *Carbon Fiber Composite Core Overhead Conductor* specifies that the bottom diameter of the tackle used for the test regarding carbon fiber composite core aluminum conductor passing the tackle is not less than 25 times of conductor diameter, and the conductor tension is 15% RTS and 20%RTS, causing the envelope angle of the conductor passing the tackle to be 20–30 degrees; the conductor makes a round trip on the tackle 12 times (namely passing the tackle 24 times) and the composite core conductor is required to have no obvious indentation or abrasion. The tackle tests below fail to strictly follow this standard, but select test parameters as needed whose severe degree exceeds standard parameters.

Ordinary single tackle is used for the tackle test and the conductor sample is carbon fiber reinforced composite core heat-resistant aluminum alloy twisted conductor JLNR/F2B-300/65 (diameter of composite core: 9 mm). The test results are as follows:

a. 660 tackle is used for the test of passing the tackle (groove bottom diameter: 660 mm). In case of envelope angle of 32 degrees and conductor tension of more than 15 kN, breakage easily happens to the composite core, as shown in Fig. 4.37.

b. 822 tackle is used for the test of passing the tackle (groove bottom diameter: 822 mm). In case of envelope angle of 45 degrees and conductor tension of more than 20 kN, breakage easily occurs to the composite core.

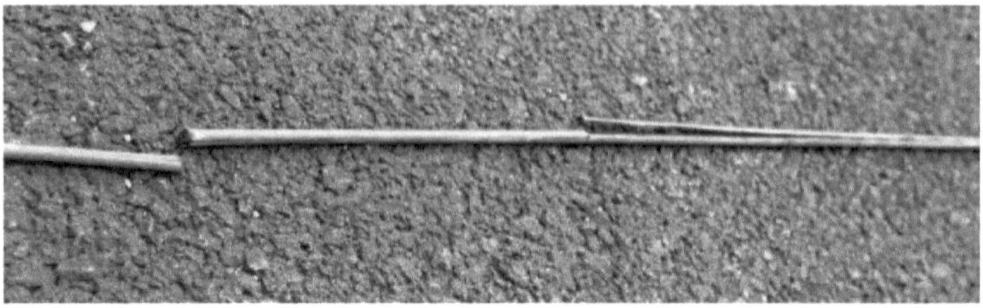

Figure 4.37 Carbon fiber reinforced composite core encountering breakage in the process of testing.

3. Double-tackle test for paying-off

   Ordinary double tackle is used for the tackle test and the conductor sample is carbon fiber reinforced composite core heat-resistant aluminum alloy twisted conductor JLNR/F2B-300/65 (diameter of composite core: 9 mm). The test results are as follows:

   a. 822 double tackle (groove bottom diameter of the tackle: 822 mm and distance between two tackle centers: 1100 mm) is used for the test of passing the tackle, the envelope angle of the conductor passing the tackle is 60 degrees, the length of the conductor sample is more than 10 m, and the conductor tension is 20 kN. The conductor makes a round trip on the tackle 20 times (namely passing the tackle 40 times), no obvious damage happens to aluminum strand, and the aluminum strand is stripped off to check whether any breakage or damage has happened to the composite core.
   b. 822 double tackle (groove bottom diameter of the pulley: 822 mm and distance between two tackle centers: 1100 mm) is used for the test of passing the tackle, the envelope angle of the conductor passing the tackle is 60 degrees, the length of the conductor sample is more than 10 m, and the conductor tension is 35 kN. A breaking sound happens to the conductor when it makes a round trip on the tackle 12 times (namely passing the tackle 24 times). Upon inspection, no obvious damage happens to the aluminum strand, however, when stripping off the staggered strand to check the composite core, it is seen that the breaking point happened to the composite core.

4. Combined tackle test for paying-off

   Combined tackle for paying-off is used for the tackle test and the conductor sample is carbon fiber reinforced composite core heat-resistant aluminum alloy twisted conductor JLNR/F2B-300/65 (diameter of composite core: 9 mm). The test result is as follows:

a. Combined tackle for paying-off is used for the test of passing the tackle. The envelope angle of the conductor passing the tackle is 100 degrees, the length of the conductor sample is more than 10 m, and conductor tension is 50 kN. The conductor makes a round trip on the tackle 20 times (namely passing the tackle 40 times); no obvious damages happen to the aluminum strand, and the aluminum strand is stripped off to check the whether there is breakage or damage to the carbon fiber reinforced composite core. In addition, no obvious displacement between conductor aluminum strand and carbon fiber reinforced composite core occurs, that is, there is no obvious core shrinking.

b. Combined tackle for paying-off is used for the test of passing the tackle. The envelope angle of the conductor passing the tackle is 145 degrees, the length of the conductor sample is more than 10 m, and conductor tension is 30 kN. The conductor makes a round trip on the tackle 20 times (namely passing the tackle for 40 times); no obvious damages happen to the aluminum strand, and the aluminum strand is stripped off to check whether there is breakage or damage to the carbon fiber reinforced composite core. In addition, no obvious displacement between conductor aluminum strand and carbon fiber reinforced composite core occurs, that is, there is no obvious core shrinking.

c. Combined tackle for paying-off is used for the test of passing the tackle. The envelope angle of the conductor passing the tackle is 145 degrees, the length of the conductor sample is more than 10 m, and conductor tension is 50 kN. The conductor makes a round trip on the tackle 20 times (namely passing the tackle 40 times); no obvious damages happen to the aluminum strand, and the aluminum strand is stripped off to check whether there is breakage or damage to the carbon fiber reinforced composite core. In addition, no obvious displacement between conductor aluminum strand and carbon fiber reinforced composite core occurs, that is, there is no obvious core shrinking.

4.5.3 Engineering Application of Combined Tackle for Paying-Off

Combined tackle for paying-off for special construction conditions has been actually applied in several projects using carbon fiber composite core

Figure 4.38 Construction site: 220 kV Hongman Line.

aluminum conductor, successfully solving the construction problem of continuous tension stringing of carbon fiber composite core aluminum conductor under special conditions.

1. Capacity expansion modification works of 220 kV Hongman carbon fiber composite core aluminum conductor in Urumchi, Xinjiang: The tackle envelope angle of No. 89 tower of the line reaches to 70 degrees, and combined tackle for paying-off is used to successfully realize continuous tension stringing (refer to Fig. 4.38).
2. Capacity expansion modification works regarding carbon fiber composite core aluminum conductor of 220 kV Yaoping Line in Yuncheng City, Shanxi: No. 41–54 section of the line has poor construction conditions and maximum envelope angle of 95 degrees, and combined tackle for paying-off is used to successfully realize continuous tension stringing (refer to Fig. 4.39).
3. Capacity expansion modification works regarding carbon fiber composite core aluminum conductor of 35 kV line from Mucun, Liulin County, Lvliang City, Shanxi to Dengjiazhuang, Liulin County, Lvliang City, Shanxi: The reverse traction conductor is needed on account of terrain limitation, the maximum envelope angle passing the tackle reaches to 160 degrees, and combined tackle for paying-off is used to successfully realize reverse traction and continuous tension stringing of carbon fiber composite core aluminum conductor (refer to Fig. 4.40).

Figure 4.39 Construction site: 220 kV Taoping Line.

Figure 4.40 Construction site: 35 kV line from Mucun, Liulin County, Lvliang City, Shanxi to Dengjiazhuang, Liulin County, Lvliang City, Shanxi.

4.5.4 Technical Features of Combined Tackle for Paying-Off

1. Combination of one group of small pulleys arranged based on some rules replaces the single big tackle, which can meet the bending radius

requirement of carbon fiber composite core aluminum conductor passing the tackle at the time of paying-off and could greatly reduce quality and dimension of the tackle. When taking combined tackle with the maximum envelope angle of 180 degrees as an example, the quality of the tackle block is only 180 kg, decreasing by 74% in comparison with that of the single big tackle with dimension decreasing to 50%.

2. Even under special construction conditions with severe terrain, the envelope angle passing the tackle of most paying-off construction is much less than 180 degrees, even less than 145 or 90 degrees; thus the feature of small pulley—easy combination can be used to design several combined tackles for conductors with different specifications and different construction requirements and make a 14-wheel tackle block suitable for the maximum envelope angle of 180 degrees, 8-wheel tackle block (refer to Fig. 4.41) suitable for the maximum envelope angle of 100 degrees, as well as a 3-wheel tackle block (refer to Fig. 4.42) suitable for the conductor with large diameter and small envelope angle passing the tackle. In this way, the dimension and quality of the tackle can be further reduced on the basis of the 14-wheel tackle block with maximum envelope angle of 180 degrees.
3. Compared with the single big tackle structure, the combined tackle has compact structure and is convenient for transportation and use.
4. Combined tackle has reasonable structure, high machining precision, and high rigidity and can ensure the straight connecting pipe and its protective steel armor will smoothly pass the tackle at the time of

Figure 4.41 8-Wheel tackle block for paying-off.

Figure 4.42 3-Wheel tackle block for paying-off.

continuous tension stringing of carbon fiber composite core aluminum conductor with connecting pipe.

5. Combined tackle is suitable for continuous tension stringing of any envelope angle and large conductor tension and greatly increase application scope of paying-off construction of carbon fiber composite core aluminum conductor tension.

Laboratory simulation test and actual engineering application indicate that compared with ordinary single tackle and double tackle, the combined tackle for paying-off has obvious technological superiority.

4.6 CONDUCTOR REPAIRING TEST

4.6.1 Technical Requirements of Conductor Repairing

In the operation process, carbon fiber composite core aluminum conductor may encounter breakage of aluminum strand and damage to carbon fiber composite core due to eolian vibration, gallop, lightning stroke, external destruction, etc., while technical features of carbon fiber composite core aluminum conductor cause its repair technology to be different from that of ordinary conductors, thus the repair index of carbon fiber composite core aluminum conductor needs to be confirmed and standardized.

For ordinary conductors such as ACSR and steel core aluminum alloy conductor, conductor repair requirements set forth in DL/T 741-2010 *Operating Standard for Overhead Transmission Line* are shown in Table 4.2.

Table 4.2 Conductor repair requirements specified by DL/T 741-2010

	Disposal method			
Type of line	Metal monofilament and pretwisted patch rods	Pretwisted patch rods and ordinary repair sleeve	Lengthened repair sleeve and pretwisted connecting strip	Straight connecting pipe, pretwisted connecting strip, and connecting pipe reinforced connecting strip
ACSR and steel core all aluminum alloy conductor	Loss of strength to the conductor incurred by the damage in the same place fails to exceed 5% of total tensile force and damage to sectional area fails to exceed 7% of the total sectional area of current-carrying part	Loss of strength to the conductor incurred by the damage in the same place fails to exceed 5%–17% of total tensile force and damage to sectional area fails to exceed 7%–25% of the total sectional area of current-carrying part	Loss of strength incurred within the scope of damage to conductor changes from 17%–50% of the total tensile force and damage to sectional area changes from 25%–60% of the total sectional area of current-carrying part	Loss of strength incurred within the scope of damage to conductor is more than 50% of total tensile force and the damage to sectional area is more than 60% of the total sectional area of current-carrying part

The mechanical strength (tensile strength) of aluminum strand of a normal conductor accounts for over 50% of overall mechanical strength of the conductor, thus the repair requirement considers not only the influence of damage to aluminum strand on conductivity, but also the influence of damage to aluminum strand on mechanical property, that is to say that due to the proportion of loss of strength incurred by the damage to aluminum strand, repair requirements for the conductor should be enhanced accordingly.

The main bearing component of carbon fiber composite core aluminum conductor is the carbon fiber composite core, and the proportion of aluminum strand mechanical strength under the condition of operation at high temperature is limited, thus the effect of the aluminum strand's mechanical strength is appropriately weakened in the process of calculating rated tensile force of carbon fiber composite core aluminum conductor.

In accordance with the above actual conditions, for the specified repair requirements for the carbon fiber composite core aluminum conductor listed by Q/GDW 1851, refer to Table 4.3.

In summary:

1. Comparing the normal conductor in Table 4.2, at the time of dividing classification of conductor repair, Table 4.3 only considers the proportion of sectional area of damage to the aluminum strand in the total sectional area of the current-carrying part, instead of the influence of damage to the aluminum strand on its mechanical strength, which is the adjustment made for the situation that the proportion of mechanical loading of aluminum strand in the carbon fiber composite core aluminum conductor is much less than that of normal conductor.
2. Q/GDW 1851: For the listed carbon fiber composite core aluminum conductor, the tensile force of carbon fiber composite core is not less than 60%, that is to say that the proportion of tensile force of aluminum strand is not more than 40%; as a result, Table 4.3 specifies that as long as the carbon fiber composite core is not damaged, it is unnecessary to use the connecting pipe to reconnect the cut conductor even in the case of complete breakage of the aluminum strand.
3. Although research and development institutions have begun research into instrument flaw detection regarding carbon fiber composite core, the degree of actual usage fails to be reached; thus, now damage to the carbon fiber composite core mainly relies on appearance inspection by the operating crew. If obvious embossment happens to the external

Table 4.3 Repair requirements regarding heat-resistant aluminum alloy twisted conductor of carbon fiber composite core

Degree of injury	Sectional area of damage to aluminum strand fails to exceed 7% of total sectional area of current-carrying part	Sectional area of damage to aluminum strand fails to exceed 7%–25% of the total sectional area of current-carrying part	Sectional area of damage to aluminum strand is not less than 25% of the total sectional area of current-carrying part	Damage to carbon fiber composite core
Disposal method	Metal monofilament and preformed armor rods patch rods	Preformed armor rods patch rods and ordinary repair sleeve	Lengthened repair sleeve and pretwisted connecting strip	Wedge joint connecting pipe, crimping connecting pipe, and pretwisted mold connection combined type connecting fittings

aluminum strand of the conductor, the screwdriver, etc. can be used to move the aluminum strand to check whether the carbon fiber composite core is damaged.

4. For the situation where the conductor needs to be cut off and reconnected in case of damage to the carbon fiber composite core, Table 4.3 lists three repair measures: wedge joint straight connecting pipe, crimping straight connecting pipe, and pretwisted wedge joint combined type connecting fittings. Of these, crimping straight connecting pipe needs to use crimping equipment for the carbon fiber composite core and external aluminum strand, wedge joint straight connecting pipe needs to use crimping equipment for the external aluminum strand, and the pretwisted connecting strip for repairing aluminum strand + wedge joint straight connecting pipe does not need to use large or heavy matched construction equipment for repairing the composite core, so its operation is the simplest and the most convenient.
5. Fig. 4.3 shows two kinds of connecting fittings of carbon fiber composite core aluminum conductor: crimping straight connecting pipe and wedge joint straight connecting pipe. The dimension of crimping connecting pipe is much less than that of the wedge joint straight connecting pipe, so it can carry out long-distance continuous tension stringing with connecting pipe and is preferentially recommended to be used in line construction. When repairing, cutting off, and reconnecting the carbon fiber composite core aluminum conductor, the combined repairing method, that is, the pretwisted connecting strip for repairing aluminum strand + wedge joint straight connecting pipe is preferred.

4.6.2 Repair Test for Broken Aluminum Strand

For repair technical requirements of carbon fiber composite core aluminum conductor, a series of verification tests below have been designed, including repair test regarding breakage of aluminum strand of carbon fiber composite core aluminum conductor and damage to composite core. Select carbon fiber composite core aluminum conductor JLNR/F2B-240/45 to conduct tension test of the repairing for broken aluminum strand. The rated tensile force of the conductor is 125.50 kN and the diameter of carbon fiber composite core is 7.5 mm.

Table 4.4 Conductor tensile force test data

Conductor model	Sample situation	Tensile force value (kN)	Average value (kN)
JLNR/F2B-240/45	In good condition	137	137
	In good condition	138	
	In good condition	136	

Table 4.5 Tensile force test result of the conductor with broken strand

Conductor model	Sample situation	Tensile force value (kN)	Average value (kN)
JLNR/F2B-240/45	13 broken strands happen to aluminum strand	110	110
	13 broken strands happen to aluminum strand	112	
	13 broken strands happen to aluminum strand	109	

1. Conductor tensile force test

   Crimp both ends of the conductor sample with the strain clamp, and then conduct the tensile force test. Refer to Table 4.4 for test results. The tensile force of three samples is more than the rated tensile force.

2. Tensile force test of conductor with broken strand

   Cut conductor aluminum strand sample into 13 strands (amounting to 25 strands), accounting for 52% of total sectional area of aluminum strand. Both ends of the conductor with broken strand are installed with the crimping strain clamp, and then the tensile force test is conducted. Refer to Table 4.5 for test results. The tensile force of three samples is more than the rated tensile force.

3. Tensile force test after repairing the conductor with broken strand

   Crimping strain clamp is installed on both ends of the conductor. Broken strands account for 52% of total sectional area of aluminum strand. MS-145 pretwisted connecting strip is used to repair the broken strand. After repair, carry out the tensile force test. Refer to Table 4.6 for test results. The tensile force of three samples is more than the rated tensile force.

Table 4.6 Tensile force test result after repairing the conductor with broken strand

Conductor model	Sample situation	Tensile force value (kN)	Average value (kN)
JLNR/F2B-240/45	After repairing 13 broken aluminum strands	136	136
	After repairing 13 broken aluminum strands	137	
	After repairing 13 broken aluminum strands	136	

Test data shows that breakage of aluminum strand can result in decline of tensile force of carbon fiber composite core aluminum conductor. However, even though sectional area of damage to aluminum strand is more than 50%, tensile force of the conductor only decreases by 19.7%, that is to say that the carbon fiber composite core is the main bearing component of the carbon fiber composite core aluminum conductor; after pretwisted connecting strip is used for repair, the conductor tension can be completely recovered.

4.6.3 The Test Repair the Damage to the Carbon Fiber Composite Core

Because carbon fiber composite core is the main bearing component of carbon fiber composite core aluminum conductor, it bears almost all conductor tension under the conditions of operation at high temperature; in case of damage to carbon fiber composite core, cutting off and reconnection is generally required to ensure mechanical strength of the conductor. Crimping straight connecting pipe, wedge joint straight connecting pipe, and pretwisted wedge joint combined type connecting fittings are used below to repair carbon fiber composite core aluminum conductor.

Cut off conductor sample, use crimping connecting pipe for crimping in the middle, install crimping strain clamp on both ends, and then conduct the tensile force test. Refer to Table 4.7 for test result.

Data indicate that after carbon fiber composite core aluminum conductor is cut off and reconnected with crimping straight connecting pipe, it can recover to the tensile force level of the original conductor.

What's more, wedge joint straight connecting pipe and pretwisted wedge joint combined type connecting fittings (repair method of pretwisted connecting stripe for repairing aluminum strand + wedge joint straight connecting pipe for cutting off and reconnecting composite core)

Table 4.7 Tensile force test data concerning repairing the conductor with crimping straight connecting pipe

Conductor model	Test sample	Tensile force value (kN)	Average value (kN)
JLNR/F2B-240/45	Crimping straight connecting pipe	139	137
	Crimping straight connecting pipe	132	
	Crimping straight connecting pipe	139	

Table 4.8 Tensile force test data concerning repairing the conductor with wedge joint straight connecting pipe

Conductor model	Test sample	Tensile force value (kN)	Average value (kN)
JLNR/F2B-240/45	Wedge joint straight connecting pipe	141	141
	Wedge joint straight connecting pipe	143	
	Wedge joint straight connecting pipe	138	

Table 4.9 Tensile force test data concerning repairing the conductor with pretwined-wedge joint combined type connecting fittings

Conductor model	Test sample	Tensile force value (kN)	Average value (kN)
JLNR/F2B-240/45	Pretwisted wedge joint combined type connecting fittings	140	140
	Pretwisted wedge joint combined type connecting fittings	141	
	Pretwisted wedge joint combined type connecting fittings	139	

can make the tensile force of carbon fiber composite core aluminum conductor effectively recover to its initial level. Refer to Tables 4.8 and 4.9 for test data, respectively.

Select one of the above three methods in the operation process according to actual conditions. It is unnecessary for pretwisted wedge joint combined type connecting fittings to use large construction equipment such as a crimping machine.

4.7 SUMMARY

This chapter emphatically introduces matched connecting fittings, and construction and operation maintenance technologies of carbon fiber composite core aluminum conductor. Compared with wedge joint connecting fittings, crimping connecting fittings applicable to carbon fiber composite core aluminum conductor substantially reduce the dimension of fittings and can be used to implement long-distance continuous tension stringing of carbon fiber composite core aluminum conductor with connecting pipe. It can used for carbon fiber composite core aluminum conductor passing the tackle based on any envelope angle to realize combined tackle for paying-off of tension stringing and provide technical support for application of carbon fiber composite core aluminum conductor under special construction conditions. The combined conductor repair method, that is, pretwisted connecting strip for repairing aluminum strand + mold connection straight connecting pipe for repairing composite core, as well as pretwisted wedge joint combined type connecting fittings can effectively relieve the repair difficulty and workload at the time of severe damage to carbon fiber composite core aluminum conductor; thus the repair technical requirements of carbon fiber composite core aluminum conductor are confirmed.

CHAPTER 5

Stress–Sag Calculation for Carbon Fiber Composite Core Aluminum Conductor

Contents

5.1 STRESS–SAG CALCULATION PRINCIPLE

Stress–sag calculation of the conductor is an important link of line design. The highest running temperature of normal aluminum conductor steel reinforced is 70°C (select 80°C by East China Grid Corporation). Within the scope of such temperature, coefficient of linear expansion, elastic modulus, and other conductor parameters can be appropriately taken as the steady state values, thus, the applied stress–sag calculation software is designed based on the premise of constant conductor parameters.

Carbon fiber composite core aluminum conductor has a migration point temperature (also called as parabolic temperature) within the scope of allowable running temperature. If the temperature is lower than migration point temperature, coefficient of linear expansion and elastic modulus of the conductor is similar to that of normal conductor. If the temperature is higher than migration point temperature, mutation happens to conductor parameters (or gradual change similar to mutation), of which conductor elastic modulus increases to the elastic modulus of carbon fiber composite core, while coefficient of linear expansion reduce greatly, and becomes less than 20% of coefficient of linear expansion at room temperature. The phenomenon pertains to coefficient of linear expansion of carbon fiber material close to zero. Aluminum strand and carbon fiber composite core bear tension together at room temperature, and conductor parameters (including elastic modulus and coefficient of linear expansion)

Engineering Energy Aluminum Conductor Composite Core (ACCC) and its Application
DOI: https://doi.org/10.1016/B978-0-12-815611-7.00005-3

are jointly determined by aluminum strand carbon fiber reinforced composite core. With the temperature rising, the elongation of aluminum strand is much higher than that of carbon fiber composite core and aluminum strand basically does not bear tension anymore, and conductor parameters (including elastic modulus and coefficient of linear expansion) are determined by carbon fiber reinforced composite core, which are actually determined by resin and carbon fiber yarn together. With the temperature continuously rising and the resin swelling, the carbon fiber yarn reaches to the state of full extension, and the conductor parameters (including elastic modulus and coefficient of linear expansion) are basically determined by carbon fiber yarn, that is to say that the final coefficient of linear expansion of carbon fiber composite core aluminum conductor is close to zero. For the situation that conductor parameters within the scope of allowable running operation of the conductor change, common stress–sag calculation software fails to be completely applicable.

Stress–sag calculation method is based on the state equation of the conductor. For the conductors hanging at fixed points on both sides, in case of the change in meteorological condition (in case of the change in air temperature and load), stress and sag of the conductor will change accordingly. In case of conductor stress σ_m, relative loan γ_m, air temperature t_m under one meteorological condition, as well as γ and t of the stress to be known under the meteorological condition, consider the elastic extension and temperature elongation of the conductor and use the principle of original line length remaining unchanged within the span under two weather conditions to list the state equation of the conductor, gain conductor stress to be known under the meteorological condition, and further calculate the conductor sag. For stress–sag calculation of carbon fiber composite core aluminum conductor, mainly select one transition state outside the known state and the state to be known and indirectly obtain stress–sag parameters under the state to be known by transition state.

In the process of the above stress–sag calculation at the temperature higher than the above migration point temperature, use a simplified method, namely assuming that parameters of carbon fiber composite core aluminum conductor (including elastic modulus and coefficient of linear expansion) are of constant values other than gradient values at the temperature higher than migration point temperature. Although the above method is not completely consistent with the actual conditions, practical application shows that this method can meet the demand of the actual engineering application.

5.2 STRESS—SAG CALCULATION EXAMPLE

Take capacity increase and transformation of one 110 kV line as an example. Calculate stress—sag of ACSR, zirconium alloy aluminum conductor invar steel reinforced, and carbon fiber composite core heat-resistant aluminum alloy twisted conductor, and summarize the technological superiority of these three conductors in light of the calculation results.

The line originally uses ACSR LGJ-240/30 with conductor safety factor of 3.04 and the maximum operating stress of 85.76 MPa. Now, intend to replace it with invar core heat-resistant aluminum alloy conductor (sectional area: 215 mm^2) or carbon fiber composite core aluminum alloy conductor JLNR/F2B-240/45 to make carrying capacity of the conductor reach to 1200 A. According to invar core heat-resistant aluminum alloy conductor parameters (refer to Tables 5.1 and 5.2), carbon fiber composite core aluminum conductor parameters (refer to Table 3.4), line meteorological parameters (refer to Table 5.3), and requirements for conductor tension (stress), calculate the stress—sag of invar core heat-resistant aluminum alloy conductor and carbon fiber composite core aluminum conductor (refer to Table 5.4).

Table 5.1 Parameters of invar core heat-resistant aluminum alloy conductor

Aluminum alloy single conductor	The number of strands/ diameter (mm)	28/3.13
	Area (mm^2)	7.694
Aluminum-clad invar	The number of strands/ diameter (mm)	7/2.75
	Area (mm^2)	5.9395
	Total	257
Computational cross-sectional area (mm^2)	Aluminum alloy	215.4
	Invar	41.6
Outer diameter (mm)		20.8
Unit length quality (kg/km)		905.8
DC resistance at 20°C (Ω/km)		0.1334
Rated stretching resistance (N)		76,600
Elastic modulus (MPa)		73,600
Linear expansion coefficient	Below migration point temperature	15.8
	Above migration point temperature	3.7

Table 5.2 Table for migration point temperature of invar conductor

Span (m)	100	150	200	250	300	350	400	450	500
Migration point temperature (°C)	77.9	84.5	89.7	93.8	97.1	99.7	101.7	103.3	104.6

Table 5.3 Line meteorological parameters

Item	Temperature (°C)	Wind speed (m/s)	Radial thickness of ice (mm)
Maximum temperature	+ 40	0	0
Minimum temperature	− 20	0	0
Annual average temperature	+ 10	0	0
Maximum wind speed	− 5	26	0
Normal icing	− 5	10	5
Operation overvoltage	+ 10	15	0
Atmospheric overvoltage	+ 15	10	0
Accident condition	+ 15	10	0
Installation condition	− 10	10	0

1. Replace it with invar core heat-resistant aluminum alloy conductor.
   a. Current-carrying capacity: The current-carrying capacity when conductor temperature is 210°C is about 1099 A.
   b. Sag: The ruling span is within the scope of 150−600 m, the conductor sag exceeds that of the original ACSR when annual average running tension and the maximum service tension of the conductor are no more than that of original ACSR, and the sag exceeds 1 m when the ruling span is more than 300 m; the sag exceeds 2 m when the ruling span is more than 400 m; and the sag nearly exceeds 3 m when the ruling span is more than 500 m. Replacing it with the invar core conductor makes it difficult to meet the requirement of safe distance.
2. Replace it with carbon fiber composite core aluminum conductor.
   a. Current-carrying capacity: The current-carrying capacity when conductor temperature is 170°C is about 1076 A.
   b. Sag: In case that the ruling span exceeds 400 m and the average running tension and the maximum running tension of the conductor are not more than that of the original ACSR, conductor sag is

Table 5.4 Comparison between stress–sag of invar core heat-resistant aluminum alloy conductor and that of carbon fiber composite core aluminum conductor

LGJ-240/30		JLNR/F2B-240/45		Invar core conductor (aluminum strand: 215 mm^2)					
Maximum temperature 70°C		Maximum temperature 170°C		Maximum temperature 210°C					
–		Migration point temperature: 110°C		Migration point temperature: 80°C		Migration point temperature: 90°C		Migration point temperature: 100°C	
Span (m)	Sag (m)	Stress (MPa)	Sag (m)	Stress (MPa)	Sag (m)	Stress (MPa)	Sag (m)	Stress (MPa)	Sag (m)
150	3.09	20.6	3.45	27.2	3.57	26.3	3.70	25.4	3.82
200	4.50	25.3	4.98	32.8	5.27	31.9	5.42	31.0	5.57
300	8.37	32.6	8.71	40.6	9.56	39.8	9.75	39.1	9.94
400	13.44	37.5	13.45	45.6	15.15	44.9	15.37	44.3	15.59
500	19.78	40.9	19.27	48.8	22.12	48.3	22.36	47.8	22.60
600	27.42	43.3	26.21	50.9	30.52	50.5	30.77	50.1	31.03

less than that of original ACSR. The conductor sag only exceeds 39–48 cm of the original steel core twisted stranded conductor when the ruling span is within the scope of 150–400 m. For the span exceeding the sag of original ACSR, line sectional view is used to directly calculate the distance of the conductor to the ground with the requirement of safe distance to the ground met.

The calculation results show that both carbon fiber composite core aluminum conductor and invar core conductor can be used for capacity increase and transformation of the line, while the advantage of carbon fiber composite core aluminum conductor is much greater than that of invar core conductor in terms of sag. The reason is that under the condition of the same aluminum section and tensile force, the quality of carbon fiber composite core aluminum conductor is obviously lower than that of the normal conductor, while the quality of invar core conductor is close to that of the normal conductor. Therefore, the invar core conductor only has the sag advantage at the high-temperature section with temperature higher than migration point temperature, while carbon fiber composite core aluminum conductor has an advantage in terms of its initial sag at the room-temperature section and its sag at the high-temperature section with temperature higher than migration point temperature; that is to say that quality advantage determines that carbon fiber composite core aluminum conductor has a greater advantage than that of invar core conductor in the capacity increase and transformation of working line.

5.3 SUMMARY

For the actual situation in which carbon fiber composite core aluminum conductor has a migration point temperature within the scope of running temperature, on the basis of normal conductor stress–sag calculation method and in the premise of meeting actual engineering application, simplify conductor parameter values at the temperature higher than migration point temperature and gain the stress–sag calculation method of carbon fiber composite core aluminum conductor.

CHAPTER 6

Standardization of Carbon Fiber Composite Core Aluminum Conductor

Contents

6.1 TECHNICAL CONDITIONS OF PRODUCT

GB/T 29324-2012 *Fiber-Reinforced Resin Composite Core Rod for Overhead Conductor* comprises the technical conditions regarding carbon fiber reinforced composite core and Q/GDW 1851 *Carbon Fiber Composite Core Overhead Conductor* comprises the technical conditions regarding carbon fiber composite core aluminum conductor, covering the current two mainstream products: heat-resistant aluminum alloy strand carbon fiber composite core aluminum conductor and soft aluminum strand carbon fiber composite core aluminum conductor.

Q/GDW 1851 *Overhead Conductor of Carbon Fiber Composite Core* is formulated mainly based on independent research and development as well as application of carbon fiber composite core aluminum conductor of the key science and technology project of the state grid corporation of china (SGCC) as well as by referring to GB/T 1179-2008 *Round Conductor Concentric Strand Overhead Conductor*, GB/T20141-2006 *Shaped Wires Concentric Stranding Overhead Conductor*, and GB/T 29324-2012 and

Engineering Energy Aluminum Conductor Composite Core (ACCC) and its Application
DOI: https://doi.org/10.1016/B978-0-12-815611-7.00006-5

GB/T 29325-2012 *Soft Aluminum Shaped Wires for Overhead Conductor*. It applies to carbon fiber composite core (single bar) overhead conductor for overhead transmission lines; specifies product classification of carbon fiber composite core overhead conductor, including carbon fiber composite core aluminum conductor products of two series, that is, heat-resistant aluminum alloy round conductor strand and soft aluminum profile wire strand; and explicitly stipulates product technical requirements, inspection rules, test methods, logo, and packaging. The main difference from normal conductor standards is that a series of technical requirements of carbon fiber composite core is added, including glass transition temperature, allowable operating temperature, coefficient of linear expansion, room temperature tensile strength, high temperature tensile strength, winding performance, torsional property, ultraviolet radiation aging performance, wet and heat aging properties, etc.

6.2 CONSTRUCTION TECHNOLOGY AND ACCEPTANCE GUIDELINE

Q/GDW 388-2009 *Construction Technology and Acceptance Guideline Regarding Carbon Fiber Aluminum Conductor Composite Core* was prepared after comprehensively considering the current situation and development trend of domestic electric transmission line and referring to construction technology guidelines for domestic extrahigh voltage and ultrahigh voltage transmission line. It specifies construction technology and acceptance standards concerning carbon fiber aluminum conductor composite core for overhead transmission line, applies to newly built and rebuilt overhead transmission line engineering with voltage class of 500 kV and above, and involves preparation before construction of carbon fiber aluminum conductor composite core, tension stringing, connection, stringing, accessory installation, and acceptance. The standard was issued and implemented in December 2009. Because it was formulated early and products and technologies related to the construction of carbon fiber composite core aluminum conductor had not yet been perfected at that time, the content of the standard has certain limitations, such as that crimping connecting fittings, pretwisted wedge joint combined type connecting fittings, long-distance continuous tension stringing of carbon fiber composite core aluminum conductor with crimping connecting fittings, and other technologies fail to be included into the standard.

6.3 GUIDELINE FOR OPERATION AND MAINTENANCE TECHNOLOGY

Enterprise standard *Guideline for Operation and Maintenance Technology of Carbon Fiber Reinforced Composite Core Heat-Resistant Aluminum Alloy Twisted Conductor*, is being formulated by SGCC with the background of full completion of the project involving independent research and development of carbon fiber composite core aluminum conductor in China. Besides carbon fiber composite core soft aluminum conductor and matched wedge joint connecting fittings introduced abroad, rich experience concerning carbon fiber composite core heat-resistant aluminum alloy twisted conductor is accumulated in the process of actual application. Several thousand meters of carbon fiber composite core heat-resistant aluminum alloy twisted conductors have been used, carbon fiber composite core heat-resistant aluminum alloy twisted conductor has been successfully applied to extrahigh voltage overhead transmission line for the first time, the technology of matched crimping connecting fittings and pretwisted wedge joint combined type connecting fittings has matured, crimping connecting fittings have been used to successfully carry out long-distance continuous tension stringing of carbon fiber composite core aluminum conductor with connecting pipe, and pretwisted wedge joint combined connecting fittings can greatly reduce repair difficulty of carbon fiber composite core aluminum conductor. Therefore, the standard has relatively comprehensive content, incorporates major achievements of the project involving independent research and development of carbon fiber composite core aluminum conductor (including product acceptance, operation standard, patrol, inspection, and maintenance of carbon fiber reinforced composite core heat-resistant aluminum alloy twisted conductor) and for the first time the standard includes installation process of wedge joint connecting fittings, and installation of crimping connecting fittings and pretwisted wedge joint combined type connecting fittings.

6.3.1 Installation Process of Wedge Joint Connecting Fittings

Wedge joint connecting fittings are one of the connecting fittings of carbon fiber composite core heat-resistant aluminum alloy twisted conductor, including straight connecting pipe and strain clamp. Carbon fiber composite core in the conductor is connected by external wedge joint and aluminum alloy strand outside the conductor is connected by crimping. Outer diameter and length of assembled straight connecting fittings are far more than that of crimping connecting fittings of the normal conductor, resulting in difficulty in passing the tackle for paying-off, so it is inappropriate to carry out long-distance continuous tension stringing with straight connecting fittings. Also, installation of such fittings needs to use hydraulic machine and other large construction equipment, so its installation is inconvenient in comparison with pretwisted fittings.

6.3.1.1 Installation Process of Straight Connecting Pipe

1. Confirm connecting pipe has all of its parts, including crimping pipe, liner pipe, wedge-shaped clamping seat, wedge clamp, and coupler.
2. Poling: Clean the surface of conductor end to be connected with cleaning cloth with length not less than three times of that of the external crimping pipe. Harness liner pipe of the straight connecting pipe from the end of the conductor and then harness external crimping pipe from the end of the conductor.
3. Make an imprint: Saw the conductor end to be even, measure the conductor with the same length at the end of conductor with wedge-shaped clamping seat, make an imprint, and twine the conductor in the conductor side of the imprint with adhesive tape to prevent broken strand, as shown in Fig. 6.1.
4. Wire stripping: Saw aluminum strand at the location of the imprint layer by layer without damaging the carbon fiber composite core. After carbon fiber composite core is exposed, wipe out oil stains on the carbon fiber composite core with dry cloth, lightly and evenly burnish carbon fiber composite core surface with special fine sandpaper, and then wipe out the powder with dry cloth as shown in Fig. 6.2. The sandpaper should not be reused.
5. Installation of wedge clamp and wedge-shaped clamping seat: Insert the carbon fiber composite core into the wedge-shaped clamping seat,

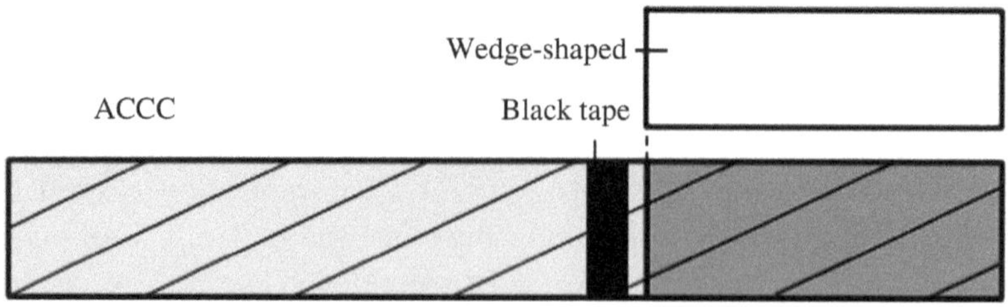

Figure 6.1 Measure the conductor with the same length by wedge-shaped clamping seat and draw an imprint.

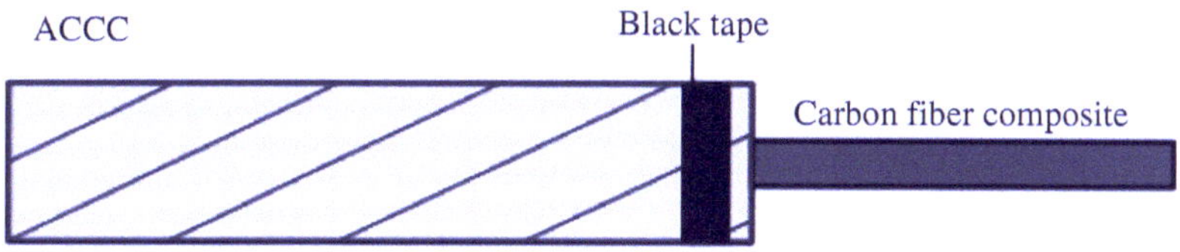

Figure 6.2 Wire stripping in the place of imprint.

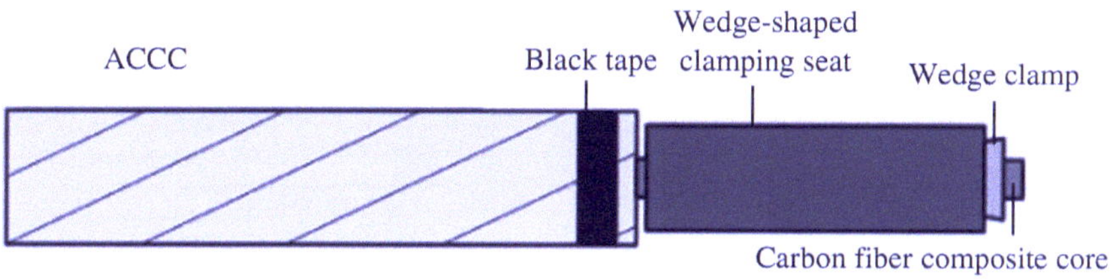

Figure 6.3 Carbon fiber composite core penetrating into wedge-shaped clamping seat and wedge clamp.

make the narrow end of the wedge clamp face the wedge-shaped clamping seat, and insert into core bar until the other end of the wedge clamp reaches 5 mm exposure from the core bar. Pull back wedge-shaped clamping seat, and make the whole wedge clamp slide into the wedge-shaped clamping seat as shown in Fig. 6.3.

6. Repeat steps 2–5 and install the conductor on another end to be connected.
7. Installation of coupler.
   a. Conduct back-to-back placement of installed conductor on both ends of the modular form clamp and wedge-shaped clamping seat. Screw the couplers into wedge-shaped clamping seat on both ends at the same time in the middle of two wedge-shaped clamping seats with both ends entering the coupler synchronously. It may be necessary to screw it in anew, as shown in Fig. 6.4.
   b. The coupler is tightened with a wrench synchronously and tightening strength shall comply with technical requirements of the product.
   c. Check 25–35 mm carbon fiber composite core exposing between aluminum strand and wedge-shaped clamping seat. Check tapered end of wedge clamp, which shall stretch out from the end of the modular form clamping seat (about 2–5 mm) as shown in Fig. 6.5.
8. Aluminum strand connection.
   a. Measure the position of the central point of the external crimping pipe with a ruler, and then make use of such position to measure

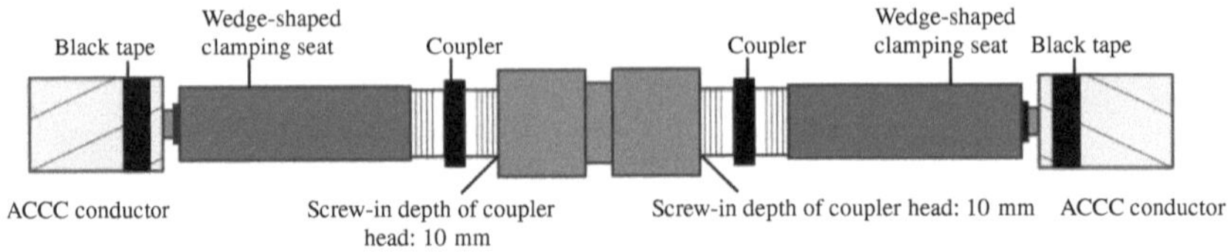

Figure 6.4 Installation of coupler.

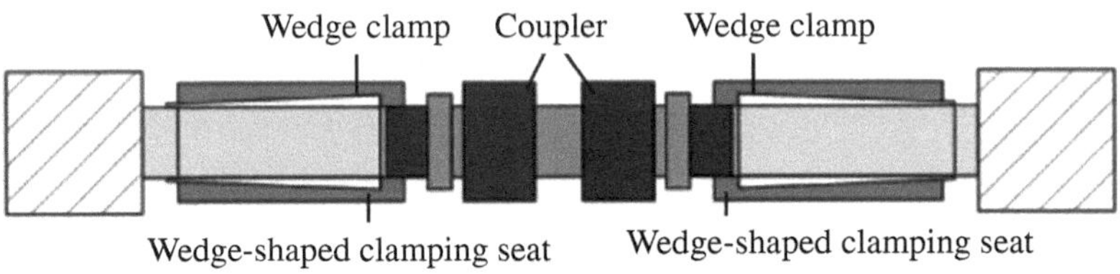

Figure 6.5 Check extension elongation of tapered end of wedge clamp.

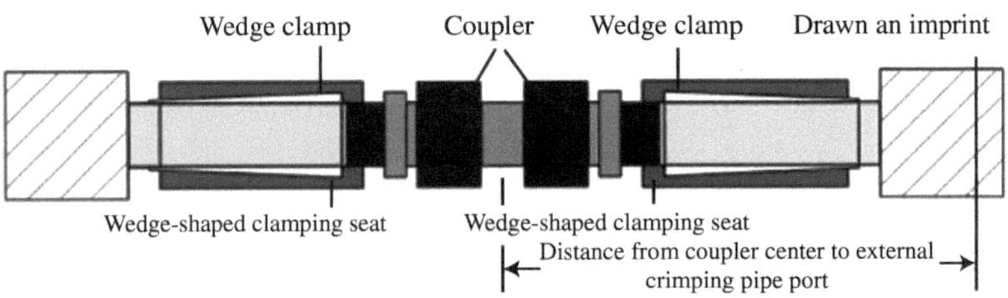

Figure 6.6 Measure the distance from external crimping pipe center to conductor end.

the distance from coupler center to ports in both sides of external crimping pipe and draw an imprint on the corresponding conductor aluminum strand in such position as shown in Fig. 6.6.

b. Evenly apply electric grease on the aluminum strand of the conductor entering liner pipe. Rub applied electric grease along the direction of aluminum strand twisting with a steel brush and then wipe out unnecessary electric grease with a clean cloth.

c. Put external crimping pipe in place based on the imprint drawn on the aluminum strand. After evenly applying electric grease on the surface of liner pipe on both sides, liner pipes on both sides are pushed into the external crimping pipe until it touches the external crimping pipe. Then put one mold in the place of 1/2 of the external crimping pipe as shown in Figs. 6.7 and 6.8.

d. Start to apply pressure to the pipe orifice end successively from the place 5 mm away from the scratch line on both ends of the

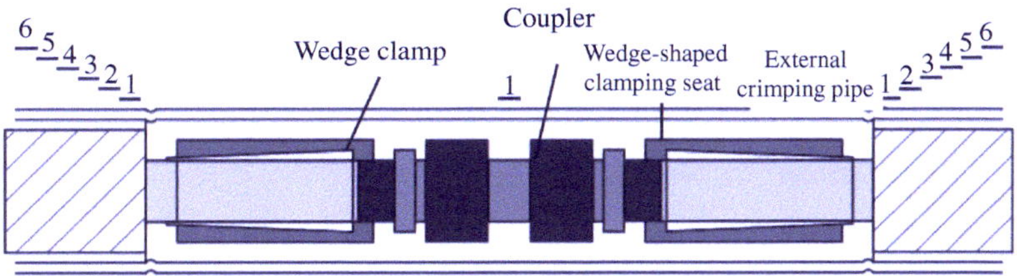

Figure 6.7 Schematic diagram 1 for pressing of connecting pipe.
Note: The number in the figure represents crimping position and sequence (the same below).

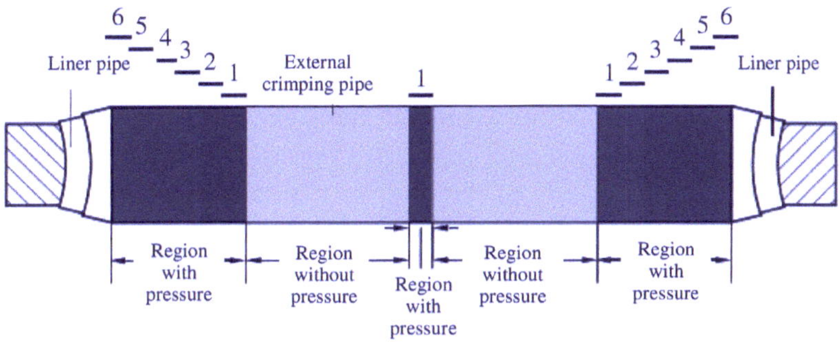

Figure 6.8 Schematic diagram 2 for pressing of connecting pipe.

external crimping pipe. While applying pressure, the overlapping region between molds shall not be less than 10 mm. See Figs. 6.7 and 6.8. Note that relative location of external crimping pipe and liner pipe in the process of crimping operation should not be changed.

e. Wipe out burrs on the external crimping pipe with a file, measure opposite side clear distance after measuring pressure with vernier caliper, measure pipe length after pressing with meter ruler, and make a record.

9. Inspection for crimping quality

a. After crimping, the curvature of the crimping pipe shall not be more than 2% of overall length of the crimping pipe.

b. After crimping, crimping pipe orifice torn conductor shall have no upheaval or loose strand and the surface of crimping pipe shall be smooth.

c. Hexagon opposite side dimension formed after crimping shall be $0.866D>$ ($D>$ outer diameter of crimping pipe). When any one of the opposite sides' dimension exceeds $0.866D > +0.2$ mm, it is necessary to replace the steel mold.

6.3.1.2 Installation Process of Strain Clamp

1. Confirm all parts of the strain clamp are present, including strain clamp connecting sleeve, coupling ring (namely steel anchor), wedge clamp, wedge-shaped clamping seat, and liner pipe.
2. Poling: Wipe out the surface of conductor end to be connected with clean cloth with length not less than three times that of the external crimping pipe, insert conductor end into liner pipe, and then penetrate into strain clamp connecting sleeve.
3. Draw an imprint: Saw the conductor end to be even, measure the conductor with the same length at the end of the conductor with wedge-shaped clamping seat, make an imprint, and twine the conductor in the conductor side of the imprint with adhesive tape to prevent broken strand, as shown in Fig. 6.1. See Fig. 6.9.
4. Wire stripping: Saw aluminum strand at the location of the imprint layer by layer without damaging the carbon fiber composite core. After carbon fiber composite core is exposed, wipe out oil stains on the carbon fiber composite core with a dry cloth, lightly and evenly burnish carbon fiber composite core surface with special fine sandpaper, and then wipe out the powder with a dry cloth as shown in Fig. 6.10. The sandpaper should not be reused.
5. Installation of wedge clamp and wedge-shaped clamping seat.
   a. Insert the carbon fiber composite core into the wedge-shaped clamping seat, make the narrow end of the wedge clamp face the wedge-shaped clamping seat, and insert into core bar until the other end of the wedge clamp reaches 5 mm exposure from the core bar. Pull back wedge-shaped clamping seat, and make the

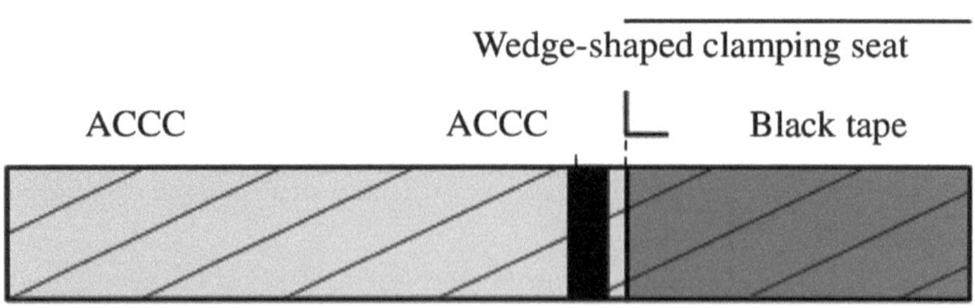

Figure 6.9 Measure the conductor with the same length by wedge-shaped clamping seat and draw an imprint.

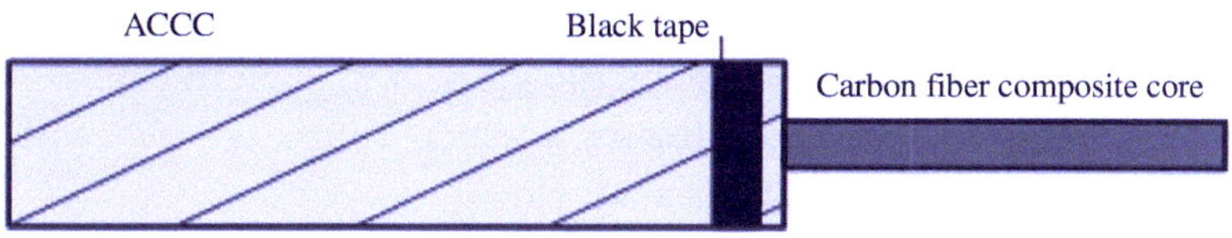

Figure 6.10 Wire stripping in the place of imprint.

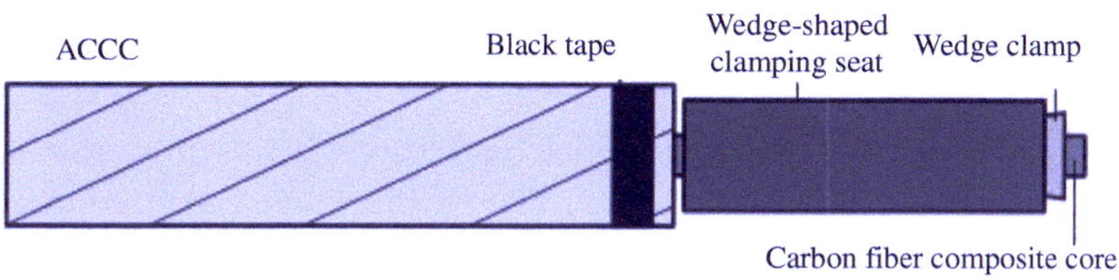

Figure 6.11 Carbon fiber composite core penetrating into wedge-shaped clamping seat and wedge clamp.

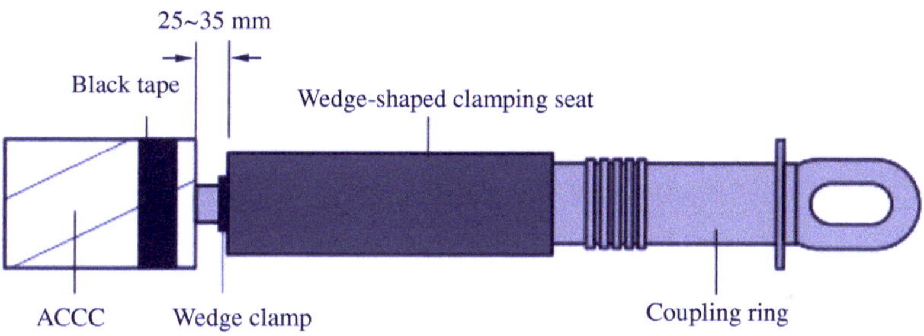

Figure 6.12 Installation of coupling ring.

whole wedge clamp slide into the wedge-shaped clamping seat as shown in Fig. 6.11.

b. Installation of coupling ring: Screw the coupling ring into modular form clamping seat and tighten it with a wrench. The tightening length shall conform to product technical requirements as shown in Fig. 6.12.

c. Check 25–35 mm carbon fiber composite core exposed between aluminum strand and wedge-shaped clamping seat. Check the tapered end of wedge clamp, which shall stretch out from the end of the modular form clamping seat (about 2–5 mm) as shown in Fig. 6.12.

6. Aluminum strand connection.

a. Evenly apply electric grease on the aluminum strand of the conductor entering the liner pipe. Rub applied electric grease along the direction of aluminum strand twisting with a steel brush and then wipe out unnecessary electric grease with a clean cloth.

b. Pull connecting sleeve of the strain clamp until it contacts with the rubber mat of the coupling ring.

c. Adjust relative angle of the coupling ring and strain clamp connecting sleeve drainage plate direction.

d. Evenly smear and brush electric grease on the surface of liner pipe and push the liner pipe into strain clamp connecting sleeve until it is withstood.

e. Apply pressure to pipe orifice end successively from the place 5 mm away from port scratch line of the strain clamp connecting sleeve conductor. While applying pressure, the overlapping region between molds shall not be less than 10 mm as shown in Fig. 6.13.

f. Wipe out burrs on the external crimping pipe with a file, measure opposite side clear distance after measuring pressure with vernier caliper, measure pipe length after pressing with meter ruler, and make a record.

7. Installation of drainage tube.

a. Remove excess electric grease in the drainage tube.

b. Measure the length of the conductor in the drainage tube at the end of the conductor and draw an imprint.

c. Remove aluminum strand oxidation film of the conductor entering the drainage tube with steel brush.

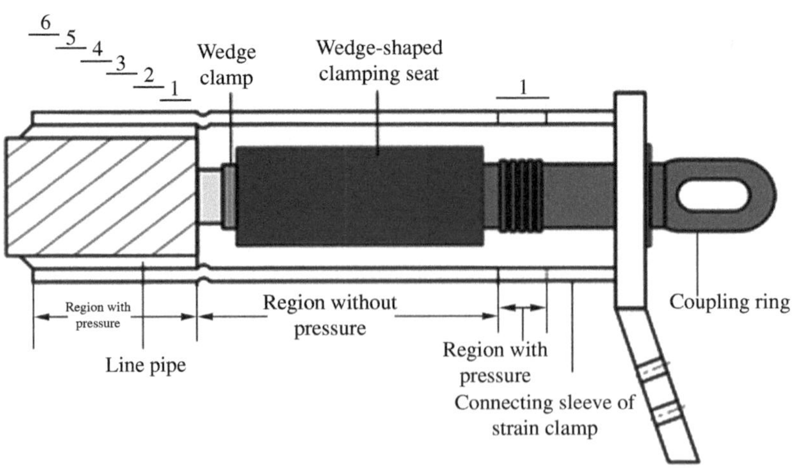

Figure 6.13 Schematic diagram for pressing of strain clamp.

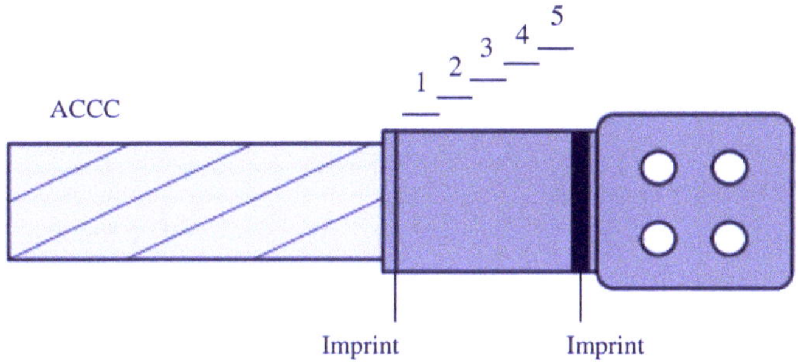

Figure 6.14 Schematic diagram for pressing of drainage tube.

   d. Make the conductor penetrate into the drainage tube whose port shall overlap with the imprint on the conductor.
   e. It is necessary to make drainage tube direction consistent with that of original bending, and apply pressure successively from the drainage tube port imprint. Refer to Fig. 6.14.

8. Inspection for crimping quality
   a. After crimping, the curvature of strain clamp shall not be more than 2% of overall length.
   b. After crimping, the conductor near strain clamp pipe orifice shall have no upheaval or loose strand and the surface of the strain clamp shall be smooth.
   c. Hexagon opposite side dimension formed after crimping shall be $0.866D$ ($D>$ outer diameter of crimping pipe). When any one of opposite sides' dimension exceeds $0.866D> +0.2$ mm, it is necessary to replace the steel mold.

6.3.2 Installation Process of Crimping Joint Connecting Fittings

Crimping joint connecting fittings are one of the connecting fittings of carbon fiber composite core heat-resistant aluminum alloy twisted conductor, and include straight connecting pipe and strain clamp. Carbon fiber composite core in the conductor and aluminum alloy strand outside the conductor shall be conducted by crimping. Outer diameter and length of assembled straight connecting fittings are close to that of crimping connecting fittings of normal conductor, resulting in success in passing the tackle for paying-off, so it is proper to carry out long-distance continuous tension stringing with straight connecting fittings. Take carbon fiber composite core heat-resistant aluminum alloy twisted conductor with diameter of carbon fiber composite core being 9.0 mm as an example below to explain the installation process of crimping connecting fittings, including crimping strain clamp and crimping straight connecting pipe.

6.3.2.1 Preparation Work

1. Preparation for hydraulic machine: Accuracy of pressure gauge of hydraulic machine shall not be less than Grade 1.5 with effective range of 2 M−100 MPa. Before startup, adjust pressure of hydraulic machine and accurately adjust displayed value by pressure gauge to comply with product technical requirements.
2. Cleaning for fittings: Before installation, clean up the contact position between steel anchor (strain clamp) or connecting steel pipe (straight connecting pipe and carbon fiber composite core) with gasoline. Wait until the gasoline in the pipe is fully evaporated before use.
3. Draw an imprint: Saw the conductor end to be even, measure crimping length $+\Delta/l$ (Δ/l refers to reservation length between steel anchor pipe orifice and aluminum strand) at the end, draw an imprint and twine conductor aluminum strand with adhesive tape at the place of imprint to prevent loose strands of conductor.
4. Wire stripping: Saw aluminum strand at the place of imprint layer by layer. At the time of cutting the internal aluminum strand, only cut it to about 3/4 of the diameter of each strand, and then break off aluminum strands one by one to prevent damage to the carbon fiber composite core.
5. Cleaning and polishing. After composite core is exposed, wipe out oil stains on the carbon fiber composite core with a dry cloth, evenly burnish carbon fiber composite core surface with special fine sandpaper, and then wipe out the powder with a dry cloth.

6.3.2.2 Poling

1. Insert strain clamp into the pipe (refer to Fig. 6.15): Insert the conductor end after pretreatment into the aluminum pipe, and then insert carbon fiber composite core with aluminum strand stripped into the steel anchor crimping zone from the steel anchor mouth until the end of carbon fiber composite core presses against the bottom of the steel anchor crimping zone. Refer to Fig. 6.15B for reservation length Δ/l between steel anchor pipe orifice and aluminum strand. At the time of passing the aluminum pipe, rotate along the stranding direction of conductor aluminum strand with length Δ/l being 3 cm.

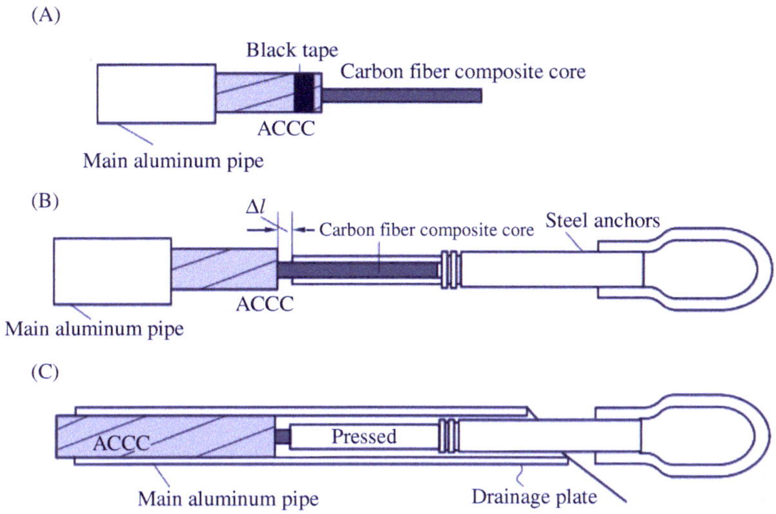

Figure 6.15 Insert strain clamp into the pipe.

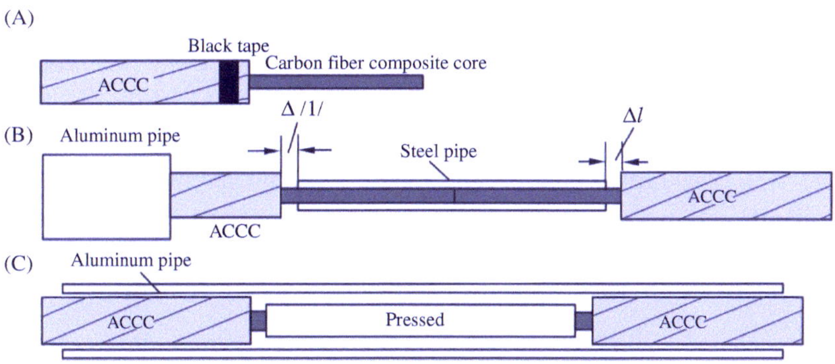

Figure 6.16 Butt-joint straight connecting pipe passing the pipe.

2. Insert straight connecting pipe into the pipe (refer to Fig. 6.16): Straight crimping connection of carbon fiber composite core is realized by abutting joint. At the time of wire stripping, length of exposed core rod is half of "steel pipe length + Δ/l," and Δ/l is 3 cm. Harness aluminum pipe into the conductor from one conductor end after pretreatment, and then make two carbon fiber composite cores with aluminum strand stripped penetrate from both ends of steel pipe until both ends of carbon fiber composite core meet each other at the mid-point in the steel pipe. The reservation length between both ends of steel pipe and aluminum strands on both sides is $\Delta/1$ as shown in Fig. 6.16B.

6.3.2.3 *Crimping*

1. Start hydraulic machine. Hydraulic operation of carbon fiber composite core shall be subject to each mold reaching the specified pressure and duration instead of the dimension after mold pressing. At the time of steel anchor or steel pipe pressing, the overlay region of two nearby molds shall not be less than 1 mm or more than 3 mm. The overlapping region length of two nearby molds at the time of aluminum pipe pressing shall conform to requirements of product technical documents. In case of no requirements of product technical documents, such length shall not be less than 1/3 of steel mold length.
2. Crimping of strain clamp (refer to Fig. 6.17).
   a. Crimping of steel anchor: When the pipe of the steel anchor crimping zone has internal diameter of 9.6–9.7 mm and external diameter of 23.8–24.1 mm, the pressure of the first five molds is 29 MPa and the pressure of the last mold is 20 MPa. Apply pressure mold by mold from steel anchor ring side to steel anchor pipe orifice side, as shown in Fig. 6.17A.
   b. Crimping of aluminum pipe: At the time the aluminum pipe is put into the pressure clamp, it is necessary to correctly adjust relative location of drainage plate and steel anchor ring. Check whether positioning imprint is in the jaw position. When closing the mold, it is necessary to make conductor and aluminum pipe level and make them consistent with the shaft center of hydraulic machine to reduce possible bending deflection after pressure is applied to

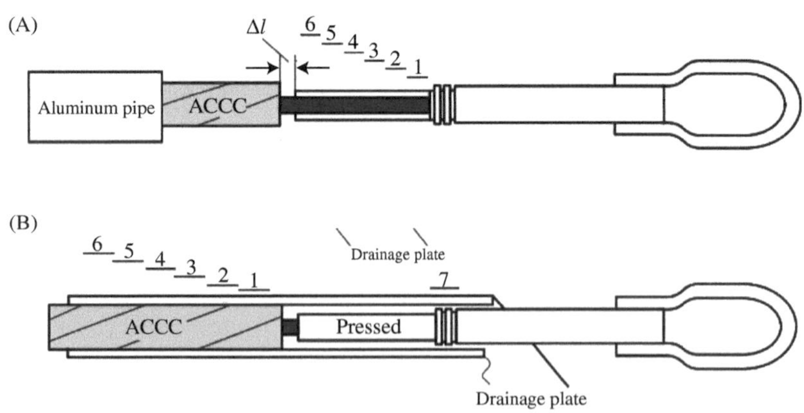

Figure 6.17 Schematic diagram for crimping operation of carbon fiber core all aluminum conductor strain clamp.

the aluminum pipe. After the position is errorless upon inspection, carry out crimping based on the sequence shown in Fig. 6.17B. Crimping of the first mold shall commence from the location of the pressure imprint on the aluminum pipe. The pressure of the first five molds is 65 MPa and the pressure of the last mold is 40 MPa. Apply pressure to one mold in reverse direction from steel anchor trough with pressure being 65 MPa.

3. Crimping of straight connecting pipe.

a. For hydraulic parts and operation sequence of connecting steel pipe of carbon fiber composite core, refer to Fig. 6.18. The center of the first mold overlaps with the center of the steel pipe; then apply pressure to pipe orifices on both sides successively and separately. When the pipe of the steel anchor crimping zone is with internal diameter of 9.6–9.7 mm and external diameter of 23.8–24.1 mm, the pressure of first five molds is 29 MPa and the pressure of the last mold is 20 MPa.

b. For hydraulic parts and operation sequence of connecting aluminum pipe of aluminum strand, refer to Fig. 6.19. Check whether pipe orifice on both ends of aluminum pipe overlaps with positioning imprint. No mold pressing happens to aluminum pipe part with steel pipe in the pipe. Apply pressure from the place with N1 pressure imprint on the aluminum pipe, and press the other side

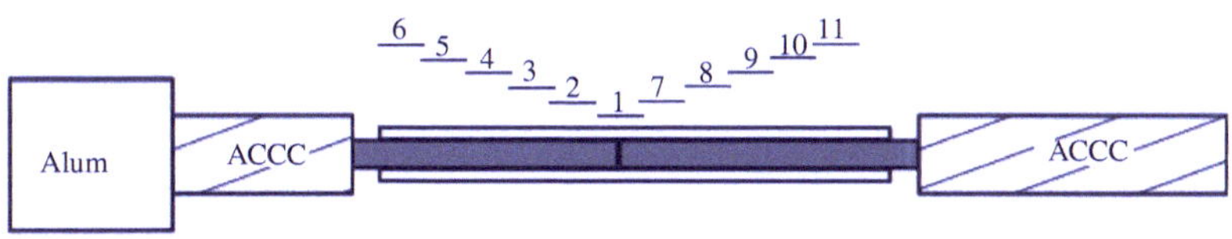

Figure 6.18 Pressing sequence of connecting steel pipe of carbon fiber composite core.

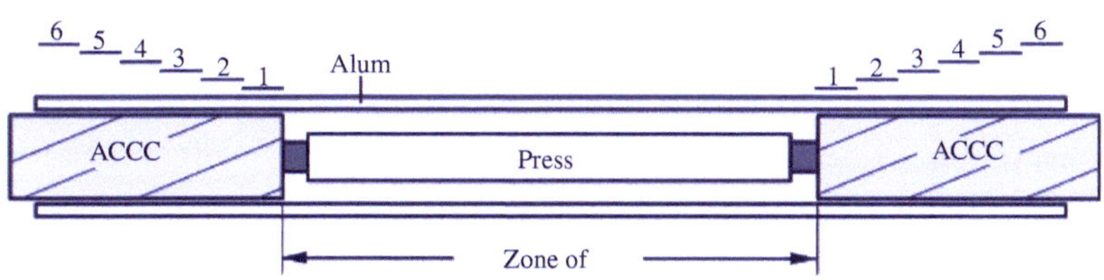

Figure 6.19 Pressing sequence of connecting aluminum pipe of carbon fiber composite core heat-resistant aluminum alloy twisted conductor.

after pressing the pipe orifice on one side. If there is no pressure imprint N1 on the aluminum pipe, measure the distance between aluminum wire ends on both sides after completion of steel pipe crimping, and then draw pressure imprint N1 on the corresponding position of the aluminum pipe. At the time of aluminum pipe crimping, the pressure of the first five molds is 65 MPa, and the pressure of the last mold is 30 MPa.

6.3.2.4 *Inspection for Crimping Quality*

1. Chamfering and removal of burrs shall be conducted for the aluminum pipe after crimping.
2. After crimping, the curvature of aluminum pipe shall not be more than 2% of overall length.
3. After crimping, the conductor near aluminum pipe orifice shall have no upheaval or loose strand. The aluminum pipe surface shall be smooth.

6.3.3 Installation Process of Pretwisted Wedge Joint Combined Type Connecting Fittings

Pretwisted wedge joint combined type connecting fittings are one of the connecting fittings of carbon fiber composite core heat-resistant aluminum alloy twisted conductor with only straight connecting fittings existing. Carbon fiber composite core in the conductor is connected by external wedge joint, and aluminum alloy strand outside the conductor is connected by pretwisted connecting strip. Internal and external connections of the connecting fittings do not need large equipment such as hydraulic machines, they are easy to install, and apply to conductor maintenance in the stage of operation and maintenance.

6.3.3.1 *Confirm That Parts of the Connecting Fittings Are Complete*

Confirm that modular form clamp, wedge clamp, coupler, and preformed armor rods (including filler strip and connecting strip) are complete.

6.3.3.2 *Connection of Carbon Fiber Composite Core*

1. Draw an imprint: Saw the conductor end to be even, measure the conductor with the same length at the end of conductor with wedge-shaped clamping seat, make an imprint, and twine the conductor in the conductor side of the imprint with adhesive tape to prevent broken strands, as shown in Fig. 6.20.

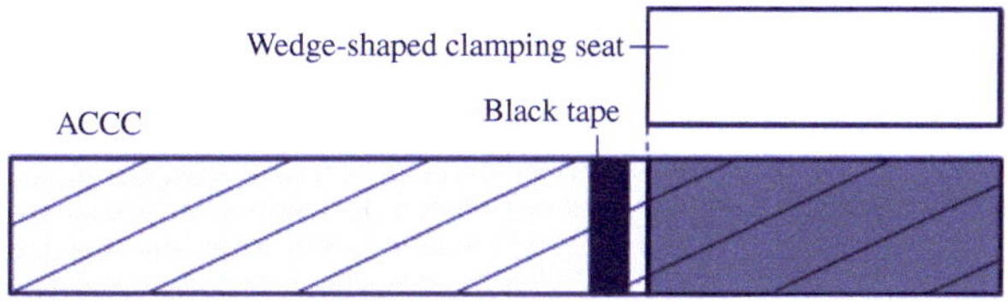

Figure 6.20 Measure the conductor with the same length by wedge clamp and draw an imprint.

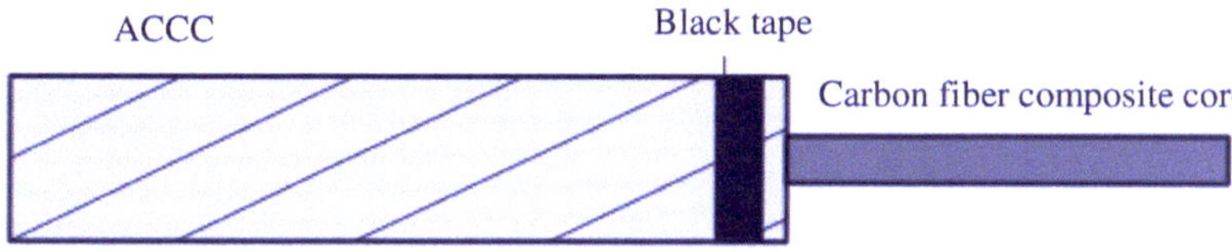

Figure 6.21 Wire stripping in the place of imprint.

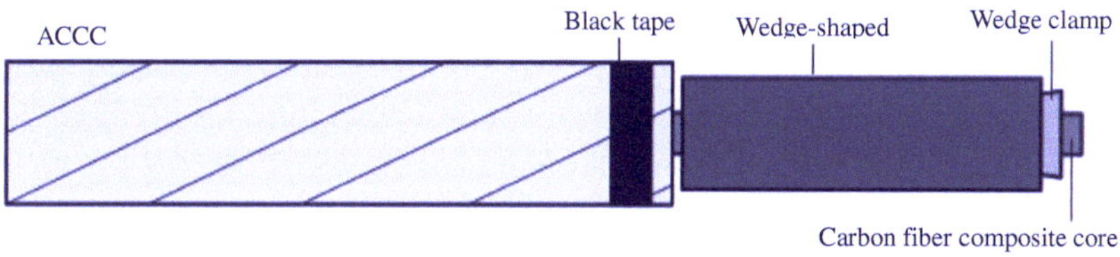

Figure 6.22 Carbon fiber composite core penetrating into wedge-shaped clamping seat and wedge clamp.

2. Wire stripping: Saw aluminum strand at the location of the imprint layer by layer without damaging the carbon fiber composite core. After composite core is exposed, wipe out oil stains on the carbon fiber composite core with a dry cloth, lightly and evenly burnish carbon fiber composite core surface with special fine sandpaper, and then wipe out the powder with a dry cloth. The sandpaper should not be reused. (Process is shown in Fig. 6.21.)
3. Installation of wedge clamp and wedge-shaped clamping seat: Insert carbon fiber composite core into wedge-shaped clamping seat, make narrow end of wedge clamp face wedge-shaped clamping seat, and penetrate into core bar until 5 mm exposure from core bar happens to another end of wedge clamp, then pull back the wedge-shaped clamping seat and make the whole wedge clamp slide into the wedge-shaped clamping seat, as shown in Fig. 6.22.
4. Repeat steps 1–3 and install the conductor on another end to be connected.

5. Installation of coupler.
   a. Conduct back-to-back placement of installed conductor on both ends of wedge clamp and wedge-shaped clamping seat. Screw coupler into wedge-shaped clamping seat on both ends in the middle of two wedge-shaped clamping seats as shown in Fig. 6.23.
   b. The coupler is tightened with a wrench synchronously and tightening strength shall comply with technical requirements of the product.
   c. Check tapered end of wedge clamp, which shall stretch out from the end of the modular form clamping seat (about 2–5 mm).

6.3.3.3 Aluminum Strand Connection

1. Installation of internal filler strip: At the time of installation, make one end of each filler strip align with the top of wedge-shaped cable clamp, conform the corresponding point of central point of filler strip on the conductor on this basis, make a mark at such corresponding point, and twine the filler strip from such corresponding point to both ends to make all lines properly fastened as shown in Fig. 6.24.
2. Installation of external connecting strip: Before installation of the external connecting strip, it is necessary to apply electric grease (also called as conductive paste) to the filler strip and all filled conductor aluminum strands, and also to the conductor based on the length of external connecting strip. Take the central point of the connecting coupler as a corresponding point of the central point of each connecting strip and make a mark at the corresponding point. Install the first

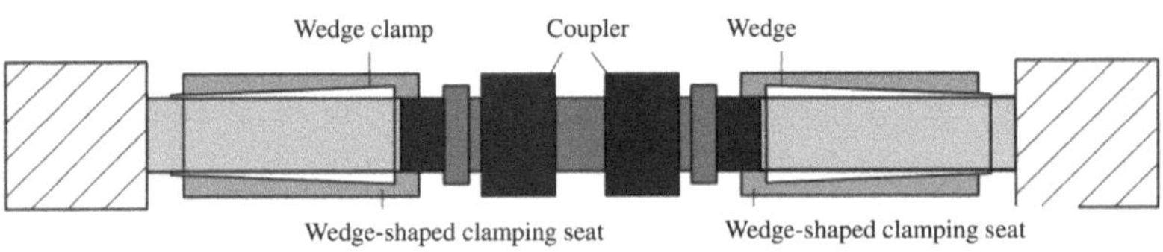

Figure 6.23 Install the coupler and check extension length of wedge clamp tapered end.

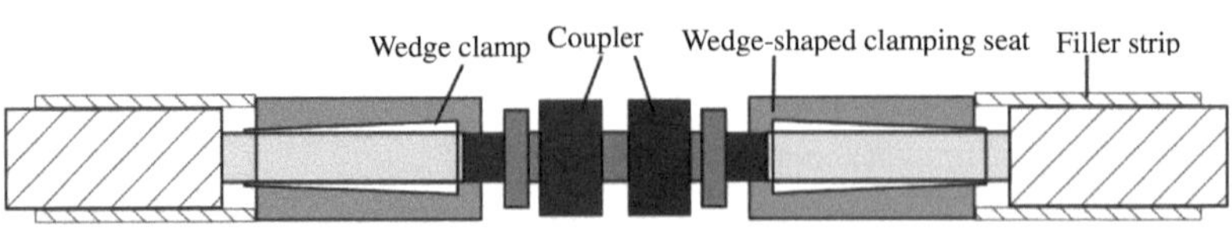

Figure 6.24 Installation of internal filler strip.

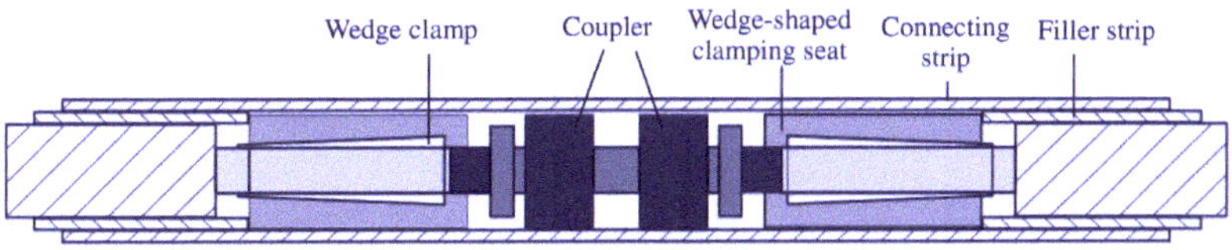

Figure 6.25 Installation of external connecting strip.

external connecting strip, align the central mark, and twine the connecting strip from the central mark to both ends, respectively; then install other external connecting strips successively and make sure each strip is properly fastened, as shown in Fig. 6.25. In the installation process, note that because only carbon fiber composite core of the conductor bears all the external forces at the time of twining external connecting strip and the flexibility of carbon fiber reinforced composite core is relatively low, it is necessary to keep the conductor horizontal in the installation process, fix the conductor, and reduce conductor torsion and bending to avoid damage to the carbon fiber reinforced composite core.

6.4 SUMMARY

A series of standards regarding carbon fiber composite core aluminum conductor formulated for the first time, including national standard *Fiber-Reinforced Resin Composite Core Rod for Overhead Conductor* and enterprise standards from SGCC, namely, *Overhead Conductor of Carbon Fiber Composite Core*, *Construction Technology and Acceptance Guideline Regarding Carbon Fiber Aluminum Conductor Composite Core*, and *Technical Guidance for Operation and Maintenance of Carbon Fiber Composite Core Heat-Resistant Aluminum Alloy Twisted Conductor*, involve technical conditions, construction, operation, maintenance, etc., of carbon fiber aluminum conductor composite core and are of great importance in fully standardizing manufacturing, construction, operation, and maintenance of carbon fiber composite core as well as promoting popularization and application of the production, updating, and rebuilding of the power grid.

CHAPTER 7

Applicable Scope of Carbon Fiber Composite Core Aluminum Conductor

Contents

Since the carbon fiber composite core aluminum conductor was put into application, it has been used and limited to capacity increase and transformation of short-distance lines. Although its technological superiority is exerted to a certain extent, its application scope is narrow and it fails to give play to social economical benefits in other fields of electric transmission line. Such situation is related to the exorbitant price of carbon fiber composite core aluminum conductor as well as its insufficiently mature construction, operation, and maintenance technologies. This section

Engineering Energy Aluminum Conductor Composite Core (ACCC) and its Application
DOI: https://doi.org/10.1016/B978-0-12-815611-7.00007-7

systematically analyzes the application scope and application models of carbon fiber composite core aluminum conductor and provides a basis for wider popularization and application of carbon fiber composite core aluminum conductor.

7.1 LINE CAPACITY EXPANSION APPLICATION

7.1.1 Capacity Expansion of Medium-Distance and Short-Distance Lines

Medium-distance and short-distance lines (especially short-distance line) are the bottleneck for transmission of electricity, mainly involving thermal stability extremity of the conductor, thus normal aluminum conductor steel reinforced (ACSR) replaced by carbon fiber composite core aluminum conductor with advantageous thermal stability can effectively improve the delivery capacity of medium-distance and short-distance lines as well as save one-time investment of power grid construction.

If normal large-section ACSR or heat-resistant aluminum alloy conductor with the same section as that of the original conductor is replaced at the original line, the former causes sag exceeding the original conductor due to increase in conductor quality, and the increase in conductor tension may exceed tower load requirements, which results in sag exceeding the original conductor due to operation of conductor at high temperature. In short, for the former or the latter, original iron tower design load and sag fail to meet the requirements of conductor replacement, necessitating replacement by a new tower that is higher and (or) has larger mechanical strength. In fact, it means building one new line at the original corridor; as a result, the power will be cut for a long time during construction, great pressure will be put on the system operation, a big outsourcing difficulty due to construction will arise, compensation costs will be increased, and construction cost will be greatly increased. Such compensation costs tend to increase year by year; various demolition and compensation costs are even equivalent to construction costs of the line.

If ACSR is replaced by carbon fiber composite core aluminum conductor with the same aluminum section and tensile force, the quality advantage (refer to Table 7.1) and sag advantage of carbon fiber composite core aluminum conductor can ensure that the capacity expansion goal can be achieved without the tower and foundation being replaced, and construction and compensation cost can be minimized. Adverse factors also exist in case of using carbon fiber composite core aluminum conductor

Table 7.1 Qualitative comparison between steel core and carbon fiber composite core

Core section (mm^2)	Tensile force (kgf)	Quantity of carbon fibers	Quality of carbon fiber core (kg/km)	Quality of steel core (kg/km)	Carbon fiber core/steel core (%)
50	1550	1550/120 = 13	10.4	62.7	17
240	7860	7860/120 = 66	52.8	432	12
240 (reinforced type)	9410	9410/120 = 79	63.2	564	11
300	8360	8360/120 = 70	56	373	15
600	16,200	16,200/120 = 135	108	741	15

for capacity increase and transformation of the line. Firstly, the conductor always needs to run at high temperature, causing line damage increases. Secondly, the conductor is expensive. Since carbon fiber composite core aluminum conductor independently researched and developed in China was put into actual application, its price has been reduced by 50%, which is still relatively higher than that of normal ACSR, and is about 3–4 times of that of ACSR. Even so, based on full-life economic comparison, namely overall consideration of one-time investment of capacity increase and transformation and loss in the operation period, especially expensive expropriation costs and other compensation costs, this method of capacity increase and transformation still has remarkable economic and social benefits. Therefore, replacement by carbon fiber composite core aluminum conductor at the current line is the preferred way for solving the above capacity-increase problem.

Taking the engineering of capacity increase and transformation for 220 kV double-line of a company as an example, the engineering is divided into two sections: double-circuit reformation section and single-circuit reformation section.

1. Technical scheme of double-circuit reformation section. At the double-circuit reformation section (I NO-I N14, IIN0-II14), Lines I and II are hung on two double circuits respectively with line length amounting to 4.3 × 2 km. Conductors of Lines I and II are 2XLGJ-240. According to line reformation requirements, two reformation schemes are proposed:

   Double-circuit section scheme 1: 2 × JLNR/F2B-240/45 carbon fiber composite core aluminum conductor is adopted, the original line tower is used, and the insulator is replaced. Comprehensive cost of the line is RMB 41.75 million.

   Double-circuit section scheme 2: 2 × LGJ-400/35 ACSR is adopted, old line of 4.3 × 2 km is removed, double circuit of 4.3 × 2 km is newly built, and the comprehensive cost of the line is RMB 60.04 million.

2. Technical scheme of single-circuit reformation section. At the single-circuit section (IN14-IN47, II14-II N50), Lines I and II are two parallel single circuits with line length of 10.2 × 2 km. Conductor of Line I is LGJQ-400 and conductor of Line II is 2 × LGJ-240. According to line reformation requirements, two reformation schemes are proposed:

   Single-circuit section scheme 1: 2 × JLNR/F2B-240/45 carbon fiber composite core aluminum conductor is adopted, original line tower is used, and the insulator is replaced. Comprehensive cost of the line is RMB 29.85 million.

Single-circuit section scheme 2: 2 × LGJ-400/35 ACSR is adopted as conductor and double circuit of 10.2 km is newly built. Comprehensive cost of the line is RMB 47.59 million.

3. Comparison for comprehensive engineering investment (refer to Table 7.2). Based on Scheme I (carbon fiber composite core aluminum conductor scheme) selected for single-circuit reformation section and Scheme 1 (carbon fiber composite core aluminum conductor scheme) selected for double-circuit reformation section, comparison for comprehensive investment of the engineering is that dynamic investment of the ACSR (LGJ-400/35) scheme is RMB 107.9808 million while dynamic investment of the carbon fiber composite core aluminum conductor (JLNR/F2B-240/45) scheme is RMB 71.9508 million with investment of RMB 36.03 million saved. The carbon fiber composite core aluminum conductor (JLNR/F2B-240/45) scheme achieves the capacity-increase objective and saves investment of about 33%, so it is the most reasonable scheme. Pursuant to calculated results of capacity increase and transformation engineering of the other carbon fiber conductor, 50%–70% of engineering investment can be saved.

7.1.2 Capacity Expansion of Long-Distance Lines

With advances in technology, advanced electric power transmission techniques such as series compensation and compact transmission line are widely used in long-distance lines, which greatly increases the transient

Table 7.2 Table for comparison of comprehensive engineering investment RMB 10,000

Dynamic investment of engineering		Double-circuit reformation part	Single-circuit reformation part	Total
ACSR	OPGW optical cable	–	35.08	35.08
(LGJ-400/35) scheme	Circuit	6004	4759	10,798.08
Carbon fiber composite core aluminum conductor	OPGW optical cable	–	35.08	35.08
(JLNR/F2B-240/45) scheme	Circuit	4175	2985	7195.08

state stability limit of long-distance transmission line, likely causing a bottleneck to the thermal stability limit of the conductor transmission of the line. Based only on technical characteristics, carbon fiber composite core aluminum conductor can be completely used for capacity increase and transformation of the above long-distance line. However, now there is a universal opinion that carbon fiber composite core aluminum conductor is only applicable to capacity expansion of medium-distance and short-distance lines around the city with dense population and high cost of land and is not applicable to long-distance lines, due to the high price of carbon fiber composite core aluminum conductor. Although ultrahigh voltage and extrahigh voltage long-distance lines always go through sparsely populated areas, resulting in relatively low land acquisition and construction compensation costs, such costs cannot completely offset high price of carbon fiber composite core aluminum conductor. However, by using the first dry-jet wet-spinning high-performance carbon fiber (T700) production line constructed and put into operation as an opportunity, domestic technology of carbon fiber composite core aluminum conductor will be fully developed, and high-performance carbon fiber and carbon fiber reinforced composite core will greatly increase so as to fully exploit price reduction potentialities of carbon fiber composite core aluminum conductor, urging the price of carbon fiber composite core aluminum conductor closer to that of normal conductor, and making it applicable to medium-distance and short-distance lines in areas with high cost of land and enabling its widespread use in capacity increase and transformation engineering of ultrahigh voltage and extrahigh voltage lines with equal distance.

7.1.3 Selection of Conductors' Sectional Area

Capacity expansion effect of carbon fiber composite core aluminum conductor is closely related to hot operation performance of the conductor. Although hot operation represents the improvement of transmission capacity, it also means an increase in transmission losses, thus in case of choosing carbon fiber composite core aluminum conductor, it is necessary to avoid small conductor section and excessively increase operation temperature of the conductor, as well as observe some sectional area selection principles.

1. Under the normal operation mode of the line, operation temperature of carbon fiber composite core aluminum conductor shall be

controlled to be relatively low (below 40°C similar to that under the state of the normal conductor). As the line runs this way most of the time, the transmission losses are made to be at a relatively low level. It is unnecessary to reduce the conductor section for embodying sag advantage and high temperature resistance advantage of carbon fiber composite core aluminum conductor, which leads to the conductor running at high temperature for a long time and causes higher transmission losses.

2. Only under the operation form of $N-1$ (individual important user or power transmission channel selects $N-2$ operation form), can carbon fiber composite core aluminum conductor exert high temperature resistance performance and run at high temperature (such as 70–150°C). $N-1$ (or $N-2$) operation mode of the line is only one kind of transient and situational approach to a certain extent (such as line tripping and line interruption maintenance). Under most normal operation modes, control conductor temperature should be less than 40°C of line loss, and it is unnecessary to adopt the conductor with large cross section to control conductor temperature under $N-1$ mode lasting for a short time to be lower, causing wasting of resources.

7.2 APPLICATION OF LONG-SPAN LINE

7.2.1 Brief Introduction of Long-Span Line

Construction of high-tension overhead line cannot avoid the crossing of rivers, straits, etc. Due to large span length of transmission lines crossing broad water surfaces and the navigation requirements of many rivers or straits, special design is required. In general, in case of span length of over 1000 mm and straight tower crossing of over 100 m, navigation requirements arise. Such a specially designed power transmission sequence is called a long-span transmission line.

Meteorology conditions for the design of long-span lines are more severe than those of ordinary lines, with higher safety requirements than those of ordinary lines as well. Life expectancy is designed based on 50 years with large quantities, long construction period, and construction cost far more than that of normal lines. Unit investment of one piece of long-span transmission line engineering is more than 10 times to dozens of times of that of ordinary transmission line. The investment of single engineering changes from RMB tens of millions to RMB hundreds of millions. Therefore, in the long-span transmission line engineering, it is a must to carry out wide, in-depth, and meticulous work, repeatedly compare various design schemes, popularize and apply safe and reliable new technology, new processes, and new technology, and realize optimal design in the premise of ensuring engineering quality to reduce materials consumption of long-span engineering, lower engineering investment, shorten construction cycle, and maximize economic benefit and social benefit.

7.2.2 Selection of Long-Span Line Conductor

Conductor characteristics determine height and performance load of crossing tower as well as transmission capacity. Selection of the proper conductor has an important significance on technical–economic indicators of long-span line engineering.

The conductor section of ordinary line shall be selected in accordance with economic current density of the conductor, while the conductor section of a long-span line can be selected pursuant to thermal stability extremity of the conductor, that is to say that transmission losses of long-span line are not the decisive factor of the design, while lowering conductor sag and reducing conductor tension (namely lowering height of crossing tower, saving tower material, lowering the basic requirement of crossing tower, lowering anchor tower at both sides and its basic requirements accordingly, and reducing the cost of the whole engineering) are factors to be emphatically considered.

Take long span of 500 kV Zhoushan as an example: The crossing tower is 370 m in height with weight of single tower being 6000 t, and the tower material is high-strength steel imported from Japan completely. The one-base common tangent tower (500 kV) is only about 10 t. In case of assuming that an average strain tower is about 30 t with average weight of tower being 20 t, the quality of the above one-base crossing tower equals to that of a 300-base common tower, equivalent to 150 km 500 kV line (span is 500 m). And because the above comparison fails to consider the difference of tower materials, carbon fiber composite core aluminum conductor with superstrong thermal stability, high strength, and lightweight theoretically has a significant advantage in the application of long-span line.

7.2.3 Calculation Case of Long-Span Line

The 1000 kV Huainan-Shanghai power transmission and distribution project is the first extrahigh voltage double-circuit alternating current

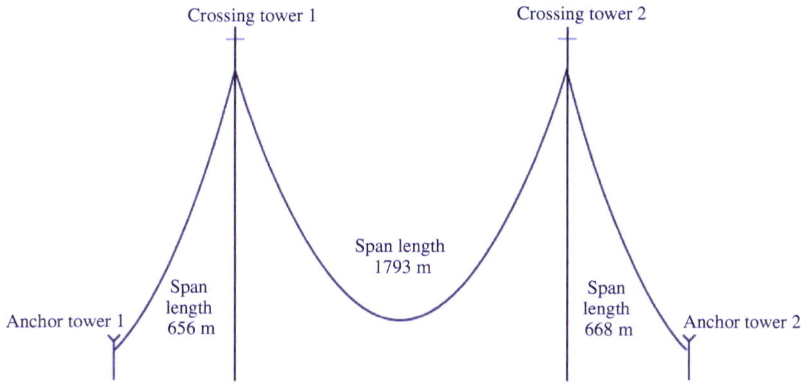

Figure 7.1 Cross-sectional drawing of 1000 kV Jiyang long-span line.

transmission line based on eastern power transmission of Anhui, which crosses the Yangtze River in Jiyang, Anhui (refer to Fig. 7.1). The section takes 1000 kV Jiyang long-span line as an example to compare stress–sag of extra strong steel core high-strength aluminum alloy twisted conductor AACSR/EST-640/290 × 6 currently proposed to be designed and used with that of carbon fiber composite core heat-resistant aluminum alloy twisted conductor JLNR/F2B-630/200 × 6 assumed to be used. The selection principle of carbon fiber composite core heat-resistant aluminum alloy twisted conductor is similar to that of extra strong steel core high-strength aluminum alloy twisted conductor with similar section of aluminum alloy strand. At the time of tensile force calculation, it is necessary to only consider the mechanical strength of the composite core. Although the conductor fails to belong to the formal product, it is reasonably fabricated in light of actual features of carbon fiber composite core aluminum conductor. Refer to Tables 7.3 and 7.4 for details. Refer to Table 7.5 for meteorological conditions.

The calculation results are shown in Table 7.6 with calculation based on average stress being $20\%\sigma_b$. When compared with extra strong steel core high-strength aluminum alloy twisted conductor, sag of carbon fiber composite core aluminum conductor declines to 78.21 m from 131.03 m, decreasing by 40%. In case of calculation based on average stress being $21\%\sigma_b$, when compared with extra strong steel core high-strength aluminum alloy twisted conductor, sag of carbon fiber composite core aluminum conductor declines to 75.28 m from 125.08 m, decreasing by 39.8%, that is to say that the tower height can decline by about 50 m, and the descent range is very considerable in comparison with tower nominal

Table 7.3 Parameters of extra strong steel core high-strength aluminum alloy twisted conductor AACSR/EST-640/290 [1]

Item		Unit	Product warranty value
Structure	Aluminum profile conductor	Pieces	42 × 4.40
	Steel core	Pieces	37 × 3.14
Nominal cross section	Aluminum profile conductor	mm^2	638.62
	Steel core	mm^2	286.50
	Comprehensive cross section	mm^2	925.12
Outer diameter		mm	39.58
Calculated mass		kg/km	4017.0
Ultimate tensile strength		kN	645.3
Elastic modulus		GPa	96.81
Linear expansion coefficient		10^{-6}/°C	16.01
20°C DC resistance (calculated value)		Ω/km	0.05268

Table 7.4 Carbon fiber composite core heat-resistant aluminum alloy twisted conductor JLNR/F2B-630/200 (actual sectional area of composite core: 250 mm^2) parameters

Item		Unit	Product warranty value
Structure	Aluminum profile conductor	Pieces	42 × 4.40
	Composite core	Pieces	1 × 18
Nominal cross section	Aluminum profile conductor	mm^2	623.45
	Composite core	mm^2	254.4
	Comprehensive cross section	mm^2	877.85
Outer diameter		mm	39.58
Calculated mass		kg/km	2164.5
Ultimate tensile strength		kN	650
Elastic modulus		GPa	96.81
Linear expansion coefficient		10^{-6}/°C	16.01
20°C DC resistance (calculated value)		Ω/km	0.04713

Table 7.5 Meteorological condition for stress–sag calculation

Calculation conditions	**Parameters**		
	Temperature, *T* (°C)	**Wind velocity, *V* (m/s)**	**Ice thickness, *C* (mm)**
Design icing	− 5	15	Conductor 15, ground 20
Maximum wind speed	− 5	33	0
Maximum temperature	40	0	0
Minimum temperature	− 15	0	0
Annual average temperature	15	0	0
Installation conditions	− 10	10	0
Lightning overvoltage	15	10	0
Operation overvoltage	15	16	0
Live work	15	10	0
Gallop situation	− 5	15	5

Table 7.6 Stress–sag calculation results (m)

Conductor model	**AACSR/EST-640/290**	**JLNR/F2B-630/200**
Sag at the time of average stress being 20%σ_b (90°C)	131.03	78.21
Sag at the time of average stress being 21%σ_b (90°C)	125.08	75.28
Nominal height of crossing tower at the time of average stress being 20%σ_b	205.03	–
Nominal height of crossing tower at the time of average stress being 21%σ_b	199.08	–

height of about 200 m. And the above calculations are based on the highest operation temperature of these two conductors being 90°C; if calculation based on the highest operation temperature of carbon fiber composite core aluminum conductor is 150°C, the advantage arising when carbon fiber composite core aluminum conductor is applied to long-span line will be more significant.

7.2.4 Problems Occurring in Long-Span Line Application

Currently, there are some technical obstacles that occur when carbon fiber composite core aluminum conductor is applied to long-span line.

Long-span line always needs the conductor with rated tensile force of over 300 kN, but the maximum composite core diameter of the carbon fiber composite core aluminum conductor that can be actually put into use now is 9.5 mm, and rated tensile force of the conductor is about 250 kN; as a result, it is difficult for the conductor to meet the application requirement of long-span line, and it thus becomes a must to research and develop carbon fiber composite core aluminum conductor with higher strength. Research and development thinking is as follows:

1. Multistrand twisted carbon fiber composite core

If single composite core is used to make high-strength or superstrength carbon fiber composite core aluminum conductor based on the current design form of carbon fiber composite core aluminum conductor, the diameter of the composite core needs to be greatly increased; as a result, winding, transportation, construction, and other properties of the conductor will decline, which goes against actual application.

To improve mechanical strength of carbon fiber composite core aluminum conductor as well as realize its construction and application properties, one kind of technological approach is to change the design form of carbon fiber composite core and twist multistrand thin composite core with twisted carbon fiber composite core as shown in Fig. 7.2. On one hand, the mechanical fatigue resistance properties of this kind of multistrand twisted carbon fiber composite core are better than that of single composite core on the same section due to the application of multicore twisting, and the breakage risk of the whole carbon fiber composite core caused by local defect of single core structure can be effectively avoided. On the other hand, the multicore structure has the small inertia moment of the carbon fiber composite core and good suppleness, can effectively reduce scroll diameter, has an obvious advantage over single carbon fiber composite core with major diameter in terms of suppleness and bending capacity, and facilitates rolling, packaging, transportation, and construction. Therefore, sectional areas of composite core can be increased by increasing the number of strands of this kind of multistrand twisted carbon fiber

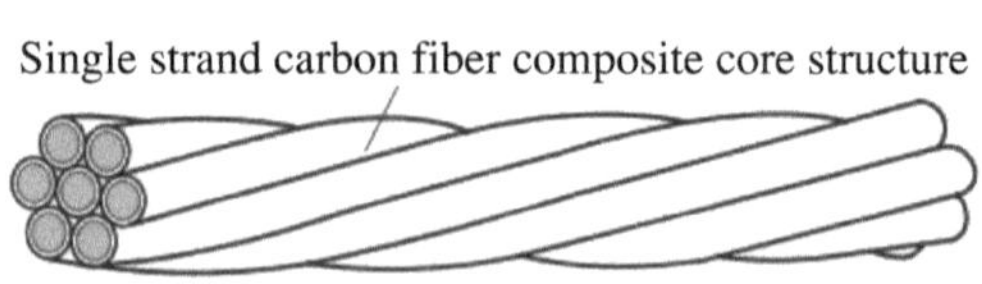

Figure 7.2 Schematic diagram of multistrand twisted carbon fiber composite core structure.

composite core, so as to substantially raise the rated tensile strength of the composite core and composite core conductor and gain a carbon fiber composite core aluminum conductor appropriate for application to a long-span line. To date, no mature products of this form of composite core and composite core conductor have been made. If multistrand twisted carbon fiber composite core technology can be gradually perfected and put into actual use, it will possibly become an effective measure for solving the problem that the carbon fiber composite core aluminum conductor cannot possess super high tensile strength and good suppleness at the same time.

2. Carbon fiber composite core aluminum conductor of Aludur strands

Another method of improving tensile strength of carbon fiber composite core aluminum conductor is to enhance the tensile strength of the aluminum strand, namely increasing the proportion of aluminum strand mechanical strength in the overall strength of the conductor. At present, carbon fiber composite core aluminum conductor based on foreign technology mainly uses aluminum strand. Soft aluminum strand has low strength, bad deformation resistivity, and low proportion in the mechanical strength of carbon fiber composite core aluminum conductor, which can be ignored basically, making it difficult for the overall strength of carbon fiber composite core aluminum conductor to be substantially raised and causing it to fail to meet the application requirements of long-span line. Carbon fiber composite core aluminum conductor researched and developed in China also adopts heat-resistant aluminum alloy strands whose mechanical strength is greatly increased, and construction, operation, and maintenance properties of carbon fiber composite core aluminum conductor are also improved, while tensile strength of ordinary heat-resistant aluminum alloy strand carbon fiber composite core aluminum conductor fails to meet the application requirements of long-span line.

Aludur is one kind of metal material that has higher strength than ordinary heat-resistant aluminum alloy and can be used to manufacture the conductor. Tensile strength of single Aludur strand reaches to above 300 MPa, which is twice that of ordinary heat-resistant aluminum alloy strand, while its conductivity is lower than that of ordinary heat-resistant aluminum alloy strand, which is 53%IACS, that is to say that transmission losses of Aludur strand are relatively larger, as shown in Table 7.7.

As mentioned before, the conductor section of ordinary line shall be selected according to economic current density of the conductor, while the conductor section of long-span line shall be chosen on the basis of thermal stability extremity of the conductor; that is to say that transmission

Table 7.7 Comparison for mechanical and electrical performance of aluminum strands made by different materials

Single aluminum conductor material	Compressive strength (MPa)	DC resistance at 20°C (Ω m/mm^2)	Conductivity (IACS)
Soft aluminum	60–95	No more than 0.027370	63%
Heat-resistant aluminum alloy	160–188	No more than 0.028735	60%
Aludur	300–325	No more than 0.032840	53%

losses of long-span line are not the decisive factor of the design, thus to make full use of Aludur's mechanical strength advantage, it can be used as an important alternative for improving overall tensile strength of carbon fiber composite core aluminum conductor. However, as elongation at break of Aludur materials is large, reaching to 4%–8%, which is much higher than that of carbon fiber composite core (1.5%–2.5%), and high elastic modulus of carbon fiber material is the primary cause of such difference, there is great difficulty in achieving good matching of high strength of carbon fiber composite core and high strength of Aludur strand. The abovementioned multistrand twisted carbon fiber composite core may be an effective measure for solving this matching problem.

7.3 IMPROVE THE MARGIN FOR SAFE OPERATION OF THE LINE

Widely using carbon fiber composite core aluminum conductor replacing normal ACSR can realize general improvement of the margin for safe operation of the line to a certain extent, including improvement of the distance from conductor to the ground, restraining conductor gallop amplitude and enhancing mechanical fatigue resistance property of the conductor. Although improvement of the margin for safe operation is welcomed by operation units, it is hard to carry out quantum chemical calculation in the design and infrastructure construction phase, that is to say that this technological superiority fails to be recognized by design and infrastructure construction departments. So far, the widespread implementation of this application is dependent on a further decline in the price of carbon fiber composite core aluminum conductor until it is close to that of the conventional conductor.

7.3.1 Assistive Measure of Suppression of Conductor Galloping

The frequent occurrence of icing galloping and tripping accidents of overhead transmission lines has become one of the main reasons for the tripping of transmission lines, posing a serious threat to the safe and stable operation of power grids. State Grid Corporation of China (SGCC) organizes production, research, and design units to carry out research on the line of antigalloping control measures; performs explorative research from the gallop mechanism, antigalloping measures and related theories, models, and field experiments, computer simulation, and other aspects; and has accumulated a large number of transmission lines galloping data and governance experience. However, the recurring tripping failures, including the tripping failures that occurred after the measures have been taken, indicate that existing antigalloping measures and products can only function for certain galloping conditions including icing conditions and wind excitation conditions such as follows: (1) A single conductor that can rotate freely and easily forms an approximately circular icing on the conductor. Therefore, under the same wind excitation conditions, it has better performance than nonuniform icing on a multisplit conductor (such as crescent icing) to inhibit gallop. Based on this principle, Japan and China have developed rotary subconductor spacers. However, this kind of approximately circular ice coating can still gallop intensely under suitable wind excitation conditions. (2) In recent years, the widely used interphase spacers have destroyed the stable vibration of the conductor by interfering with each other. However, during the heavy snowstorm in North China in November 2012, there were no interphase spacer antigalloping measures in various arrangements to avoid galloping and tripping.

The above situation is not intended to negate the existing antigalloping measures and products. However, we should also avoid exaggerating the effects of certain kinds of products and measures. The idea of antigalloping in the serious galloping areas may emphasize the combination of various antigalloping measures such as: At the same time the rotary subconductor spacers and interphase spacers are installed, and light conductor and so on are used.

Under the condition of galloping, the amplitude of the conductor can be estimated as follows:

$$a = \eta \frac{2vl \sin \beta}{n} \sqrt{\frac{m}{T}}$$

where α is single peak of conductor galloping (m); ν, conductor heart velocity (m/s); l, span (m); T, conductor tension (N); m, the mass per unit length of the conductor (N/m); η, galloping meteorological coefficient, generally $\eta = 0.15$; β, the angle between wind direction and the axis of the conductor; n, half-wave number, that is, the order of galloping. When the span $l < 200$ m, n may be 1 or 2; $200 \leq$ span $l < 600$ m, n may be 2 or 3; span $l \geq 600$ m, n may be 3 or 4.

The above galloping model shows that the galloping amplitude of the conductor after galloping is related to the parameters of the conductor tension and mass, and the galloping amplitude decreases as the conductor tension increases, and decreases as the conductor mass decreases. Compared with the conventional conductor, carbon fiber composite core aluminum conductor is a light conductor, with significant mass advantages; and the tensile strength of carbon fiber composite core aluminum conductor of the same size is much higher than the conventional conductor; that is, the margin of mechanical strength is great. Therefore, the design of the line only considers the constraints of maximum operating tension of the conductor and basically regardless of the limit of average operating tension, so that in most operating conditions, the actual tension of carbon fiber composite core aluminum conductor can be larger than the conventional conductor. Therefore, from the perspective of reducing the conductor mass and increasing the conductor tension, the carbon fiber composite core aluminum conductor has certain advantages in suppressing the galloping amplitude.

The use of carbon fiber composite core aluminum conductors as a means of suppressing galloping is even more significant for compact lines. A single conductor of a compact line has a smaller aluminum cross section than conventional lines (e.g., 500 kV compact line conductor aluminum cross section is generally 240 or 300 mm^2, and 500 kV conventional line conductor aluminum cross section is generally greater than 400 mm^2), so the use of ACSR with the use of carbon fiber composite core aluminum conductor has relatively greater difference in mass; that is, carbon fiber composite core aluminum conductor has more pronounced advantages in mass and sag, and the effect of suppressing the galloping amplitude should be better.

Carbon fiber composite core aluminum conductor as a light conductor for line galloping governance should follow the following principles: (1) The use of carbon fiber composite core aluminum conductor should first be to meet the line capacity increase or increase the safety distance between the line and the ground to prevent construction vehicles from touching the line, etc.; also, inhibiting the galloping amplitude is only suitable as an ancillary function of carbon fiber composite core aluminum conductor, so if the only intended purpose of using carbon fiber composite core aluminum conductor is to suppress the galloping amplitude, its use is not recommended. (2) Only a measure of carbon fiber composite core aluminum conductor is difficult to completely control the line galloping, so the carbon fiber composite core aluminum conductor and other antigalloping measures are used in combination, so as to achieve comprehensive management results.

7.3.2 Suppression of Wind Deflection, Construction Vehicles Touching the Line, and Other Faults

In recent years, wind deflection and other faults occur from time to time, and external accidents such as construction vehicles hitting the line have also become the main form of line fault. The carbon fiber composite core aluminum conductor has the advantages of lightweight; has obvious arc sag advantage compared with the conventional conductor; the distance between the conductor and the ground and the distance to the trees, vehicles, and buildings under the conductor are effectively increased; and when the wind deviates, the distance to the trees, buildings, and mountain slopes is effectively increased. Therefore, the application of carbon fiber composite core aluminum conductor can reduce the probability of occurrence of wind-induced discharge of the conductor, construction vehicles hitting the line, etc., and improve the ability of the line to withstand accidents.

The following calculation case illustrates the application of carbon fiber composite core aluminum conductor to improve the line margin of safe operation.

1. The relationship between conductor height and touching the line under external force: In recent years, the frequent occurrence of construction vehicles/machinery collision problems with the line is mainly due to the construction vehicle/mechanical arm directly touching the line or distance with the conductor being less than the required insulation distance, so as to form discharge channel of conductor—vehicle/mechanical—earth, resulting in line discharge tripping, and such tripping is often both unsuccessful and prone to conductor damage, broken strands, or even broken line. Increasing the height of the conductor and increasing the distance between the conductor and the

ground and the distance to the construction vehicle/machine under the conductor are the most direct and effective measures to reduce the fault of hitting the line. The distance between the conductor and the ground is related to tower call height, span, the wire mass per unit length, and other factors.

In the GB 50545-2010 *Code for Design of 110 kV–750 kV Overhead Transmission Line*, distance of 110–220 kV line to ground is considered according to vehicle cargo height of 4.0 m in the automobile transport specified by the traffic department, but the vehicle working height when collisions occur generally reaches above 7.0 m. According to the statistical data of Shanxi Electric Power Company, the average height of the 110 kV line conductors with a fault hitting the line is 8.9 m. The average calling height of the tower on both sides of the fault line is 18 m. It is recommended that 110 kV lines increase two to three times call height (i.e., 6–9 m), to make the distance of 110 kV line to ground up to 15–18 m. The average height of 220 kV line conductor with fault hitting the line is 12.4 m, and the average call height of the towers on both sides of the fault span is 24 m, so it is recommended to add 1–2 call heights to the 220 kV line (i.e., 3–6 m) to make the distance of 220 kV line to ground reach 15–18 m. The reason why 15–18 m distance from conductor to ground is proposed is that, on one hand, the distance to the ground reaches or exceeds the average working height of cranes, dumpers, excavators, and trucks, which can effectively reduce their fault of hitting the line, which accounts for 68.1% all line collision failures; on the other hand, the distance to the ground also reaches or exceeds the average distance of 15 m of the 500 kV line to ground at the point of touch-line failure, while the failure of the 500 kV line only accounts for 0.63%, indicating that increasing the distance between the line and the ground of the 110–220 kV line conductor to 15–18 m is an effective anticollision measure.

2. The light conductor is used to increase the distance from the conductor to the ground: Although the purpose of preventing the construction vehicles from reaching the line by raising the tower to improve the distance between the conductor and the ground can be achieved, the ascending tower transformation of the operation line should completely remove the original lines and reestablish the foundation, tower, and overhead lines, and must also consider offline compensation and other costs, so the total investment is often unbearable. If the light conductor is used instead of the conventional conductor to reduce sag and achieve the purpose of increasing the distance between

the conductor and the ground, the cost of disassembling the original line and rebuilding the base and tower can be saved and the channel compensation can be greatly reduced. Therefore, the cost performance of lightweight conductors such as carbon fiber composite core aluminum conductors will be significantly better than that of the ascending tower.

Taking an actual line as an example, the effect of replacing the ACSR with carbon fiber composite core aluminum conductors is compared. Table 7.8 shows the design meteorological conditions of the line. The original line uses LGJ-240/30 type conductor (see parameters in Table 7.9), intended to replace the conductor with JLNR/F2B-240/45 carbon fiber composite core aluminum conductor (see Table 3.5 parameters), and aluminum section of carbon fiber composite core aluminum

Table 7.8 Meteorological conditions

Item	Temperature (°C)	Wind speed (m/s)	Radial thickness of ice (mm)
Maximum temperature	+ 40	0	0
Minimum temperature	− 20	0	0
Annual average temperature	+ 10	0	0
Maximum wind speed	− 5	26	0
Normal icing	− 5	10	5
Operation overvoltage	+ 10	15	0
Atmospheric overvoltage	+ 15	10	0
Accident condition	+ 15	10	0
Installation condition	− 10	10	0

Table 7.9 LGJ-240/30 type conductor parameters

Steel core	Diameter	mm	7.2
	Cross section	mm^2	31.67
Comprehensive cross section		mm^2	275.96
Conductor outer diameter		mm	21.6
DC resistance of stranded conductor		Ω/km	0.1181
Weight per unit length of stranded conductor		kg/km	922.2
Calculated tensile force		kN	75.62
Elastic modulus		GPa	73.0
Coefficient of linear expansion		$°C^{-1}$	19.6×10^{-6}

conductor is the same as ACSR. ACSR maximum operating tension coefficient is 2.5, and the average operating tension coefficient is 4.0; maximum operating tension of carbon fiber composite core aluminum conductor is not greater than the original LJG-240/30 type conductor, and the average operating tension coefficient is taken as 4.0.

Table 7.10 compares the stress−sag results for both types of conductor. According to the conditions when the external force hits the line, construction operation touching the line occurs mainly from March to November every year. The coldest months of December to February lack construction, and touch-line fault can be ignored, so the focus is on the conductor lifting effect at the average temperature and maximum temperature.

Statistics data show that the average conductor height of the 110 kV line that encounters the touch-line fault is 8.9 m, and the average call height of the tower on both sides of the fault span is 18 m, which is equivalent to an average of 9.1 m of the sag; the average conductor height of the 220 kV line that encounters the touch-line fault is 12.4 m, and the average call height of the tower on both sides of the fault span is 24 m, which is equivalent to an average of 11.6 m of the sag. The sag is generally corresponding to the 350−450 m span in Table 7.10, and the corresponding sag decreases in the range of 2.60−4.16 m, which is equivalent to about one hoist (3 m) raised by the conductor. In accordance with the aforementioned requirements wherein 110 kV lines need to raise the conductor by 2−3 call heights, the use of carbon fiber composite core aluminum conductor to transform the operation of the line is still inadequate. However, in accordance with the requirements that 220 kV line conductor be hoisted 1−2 calls, carbon fiber composite core aluminum conductor can be used to transform the operation of the line.

7.3.3 Improve the Conductor Performance to Withstand Mechanical Fatigue

As mentioned above, the residual strength of carbon fiber composite core can still reach 90% of the initial strength after the tension−tension stress fatigue test. Under the same test conditions, the residual strength of steel is only 30%−40% of the initial strength. In addition, the carbon fiber composite core can successfully pass the 12×10^7 aeolian vibration fatigue test, 3×10^7 split conductor vertical vibration test, 3×10^7 split conductor horizontal vibration fatigue test, and 3×10^7 split conductor torsional vibration fatigue test; carbon fiber composite core aluminum conductor

Table 7.10 Stress–sag contrast for carbon fiber composite core aluminum conductor and ACSR

Working conditions	Span (m)	Temperature (°C)	LGJ-240/30		ACCC-240/35		Sag difference
			Stress (MPa)	Sag (m)	Stress (MPa)	Sag (m)	LGJ-ACCC (m)
Minimum temperature	200	− 20	97.4	1.681449	109.6	1.150028	0.53
	250	− 20	92.0	2.77956	109.6	1.796918	0.98
	300	− 20	87.3	4.217893	106.5	2.662485	1.56
	350	− 20	83.5	6.002831	101.7	3.797715	2.20
	400	− 20	80.6	8.12784	97.1	5.191575	2.94
	450	− 20	78.3	10.58282	93.1	6.8526	3.73
	500	− 20	76.6	13.35886	89.7	8.780592	4.58
	550	− 20	75.3	16.44935	86.9	10.970 2	5.48
Maximum temperature	200	40	50.0	3.275505	60.3	2.091487	1.19
	250	40	53.5	4.785465	64.4	3.05825	1.73
	300	40	56.1	6.561 678	66.3	4.275367	2.28
	350	40	58.2	8.610 268	67.0	5.76255	2.85
	400	40	59.9	10.936 67	67.5	7.466245	3.47
	450	40	61.2	13.54537	68.0	9.387316	4.16
	500	40	62.3	16.439 9	68.3	11.526 66	4.91
	550	40	63.1	19.62303	68.7	13.885 1	5.73
Annual average temperature	200	10	68.5	2.390257	82.5	1.528044	0.86
	250	10	68.5	3.734776	84.4	2.333046	1.40
	300	10	68.5	5.378077	83.8	3.383733	2.00
	350	10	68.5	7.320161	81.9	4.715532	2.60
	400	10	68.5	9.561026	80.2	6.28853	3.27
	450	10	68.5	12.10067	78.8	8.10189	4.00
	500	10	68.5	14.9391	77.6	10.15326	4.79
	550	10	68.5	18.07632	76.6	12.43994	5.64

has performance not less than and even better than the mechanical fatigue resistance of conventional conductor. Considering from the perspective of mechanical fatigue resistance, it can be used for the long-term safe operation of overhead transmission lines, which is in line with the good self-damping properties of carbon fiber composite core.

7.4 REDUCE THE LINE RUNNING LOSS

Under the premise that the carbon fiber composite core aluminum conductor price can be further lowered (Table 7.11), and without caring about the low mechanical strength and deformation, the fact that it is easy to damage, and other shortcomings of the soft aluminum, the general use of carbon fiber composite core aluminum strand instead of conventional ACSR of equal mechanical strength, outer diameter, and other mass can achieve a general reduction of line losses and have overall economic benefits. The reduction of this line loss and the overall economic benefits can be quantified and compared. The following analysis shows the economic benefit of carbon fiber composite core aluminum conductor to reduce the line loss.

The Nari Company proposes the following method to calculate and compare the economy of loss reduction of the carbon fiber composite core soft aluminum strand. The tensile strength of the carbon fiber composite core is more than twice that of the steel core, so under the premise of equal mechanical strength, the mass of the carbon fiber composite core is less than 20% of that of the steel core, and the outer diameter of the composite core is also smaller than that of the steel core. The mass and cross-sectional area saved by the composite core can be used to increase the conductive cross section and the mass of the aluminum strands, so that the final mass per unit length and the outer diameter of the soft aluminum strand of the carbon fiber composite core are consistent with those of the

Table 7.11 Future reasonable price of carbon fiber composite core aluminum conductor

S/N	Cross section S of conductor	Price ratio of carbon fiber conductor and the common conductor
1	$S \leq 300\ mm^2$	2.5
2	$300\ mm^2\ S \leq 630\ mm^2$	2
3	$5\ 630\ mm^2$	1.8

ordinary ACSR. With such a carbon fiber composite core soft aluminum strand instead of ACSR, conductor mass, wind load, and ice load remain unchanged, then the tower and the foundation suitable for the original ACSR are bound to apply to the carbon fiber composite core soft aluminum strand. Therefore, when calculating and comparing the economical efficiency of the two kinds of conductors, it is only necessary to compare the one-time investment difference of the conductors and the difference of the conductor loss during the service life. The investment of carbon fiber composite core soft aluminum strands is obviously larger than that of ACSR, but the conductivity of the soft aluminum strands is greater than the hard aluminum strands of the ACSR and the conductive cross section of the soft aluminum strands is larger than that of hard aluminum strands of the ACSR, so under the same transmission capacity, carbon fiber composite core soft aluminum strand loss is less than ACSR. If during the service life the loss saved by carbon fiber composite core soft aluminum strands is more than the one-time investment difference of two conductors, the use of carbon fiber composite core soft aluminum strand has better economy. Table 7.12 is the line basic operating parameters for line loss comparison, Table 7.13 compares parameters and loss of the two types of conductors for 500 kV line, Table 7.14 compares project cost of the two types of conductor for 500 kV line, and Tables 7.15–7.17, respectively, are economic comparison data for the 500, 220, 110 kV lines using carbon fiber composite core soft aluminum strands instead of ACSR. Of these, 500 kV line only takes 4.3 years to achieve a flat economy at the fastest, the 220 kV line only takes 4.8 years to achieve a flat economy at the fastest, 110 kV line only takes 6.6 years to achieve a flat economy at the fastest, while power transmission engineering design life is 30 years.

Table 7.12 Line basic operating parameters for line loss comparison

S/N	Line parameters	S/N	Line parameters
1	Single circuit 20 km	5	Loss hours 2200 h
2	Line power factor of 0.95	6	Economic current density 1.0 A/mm^2
3	The ambient temperature of the line is 15°C	7	Price ratio of carbon fiber conductor and the common conductor 1.8
4	Conductor outer diameter and mass per unit length are close		

Table 7.13 Comparison of two conductor parameters and loss for 500 kV transmission line

Conductor	LGJ-400/ 35 ACSR	ACCC-470/40 type carbon fiber conductor
Aluminum strand cross section (mm^2)	391	470
Outer diameter (mm)	26.82	26.82
Rated tensile force (kN)	103.9	110.4
Unit length quality (kg/km)	1349	1371
DC resistance at 20°C (Ω/km)	0.0739	0.0593
Line loss for first year (MWh)	11,002	8763
Electricity price (RMB/MWh)	400	
Reduced line loss for first year (MWh)	2239	
Reduced line loss percentage for first year	20%	
Reduced line loss value for first year (RMB 10,000)	89.56	
Reduced line loss per unit length of conductor (RMB/m)	3.7	
Reduced line loss value for 30 years (RMB 10,000)	2686.8	

Table 7.14 Comparison of project cost of two conductor lines for 500 kV transmission line

Conductor	ACSR 4XLGJ-400/35	Carbon fiber core conductor 4 × ACCC-470/40
Conductor (RMB 10,000)	557.28	1003.1
Reduce line loss (RMB 10,000/year)	0	89.56
Recycle period (years)	5.0	

Table 7.15 Recycling life calculation results for 500 V line using ACCC-470/4 instead of LGJ-400/35

Recycle period (years)	Transmission capacity	
Annual loss hours (h)	**1500**	**1725**
1400 h (annual maximum usable hours 3000 h)	7.8	6.8
2200 h (annual maximum usable hours 4000 h)	5.0	4.3

Table 7.16 Recycling life calculation results for 220 kV line replacing the LGJ-300/40 with ACCC-380/45

Recycle period (years)	Transmission capacity	
Annual loss hours (h)	**246.8**	**314.8**
1400 h (annual maximum usable hours 3000 h)	12.9	7.5
2200 h (annual maximum usable hours 4000 h)	8.2	4.8

Because the calculation of the project did not consider the annual interest

Table 7.17 Recovery life calculation for the 110 kV line using the ACCC-315/40 instead of the LGJ-240/35s

Recycle period (years)	Transmission capacity	
Annual loss hours (h)	**50.7**	**64.8**
1400 h (annual maximum usable hours 3000 h)	17.8	10.4
2200 h (annual maximum usable hours 4000 h)	11.3	6.6

income and other factors, and the data rigor is yet to be confirmed, this data still provides to analyze the effects of carbon fiber composite core aluminum conductor from an economic standpoint.

7.5 NEED FOR R&D OF CARBON FIBER COMPOSITE CORE ALUMINUM CONDUCTOR AT LOW PRICE

Price is one of the key factors impeding the comprehensive promotion and application of carbon fiber composite core aluminum conductor. The successful research into and development of high-performance carbon fiber and carbon fiber composite core aluminum conductor in China has broken the technology monopoly of developed countries. As the domestic high-performance carbon fiber materials and carbon fiber composite core aluminum conductor technology matures and production increases, the current price of the product has been reduced from 7 to 10 times of the conventional conductor to 3 to 4 times; even so, the carbon fiber composite core aluminum conductor still does not have conditions for completely replacing the conventional ACSR, so the need is to further search for a lower cost carbon fiber composite core aluminum conductor. Some manufacturers have proposed the ultimate price of carbon fiber composite core aluminum conductor is 1.8 times that of the conventional conductor.

Carbon fiber composite core aluminum conductor has a coefficient of linear expansion close to zero (linear expansion coefficient above the migration point temperature is less than 20% that of conventional conductor), that is, under high temperature and capacity expansion, the conductor elongation is small and the amount of sag is small. However, the vast majority of overhead transmission lines do not require operation at high temperature, that is, "coefficient of linear expansion near zero" in most cases is not meaningful, and "lightweight" of carbon fiber composite core can directly bring the advantages of reducing conductor sag and increasing safety margin under normal temperature operating conditions, which is of universal significance for overhead lines. Therefore, in a sense, "lightweight" should be the biggest technical advantage of carbon fiber composite core aluminum conductor. Under the guidance of this principle, the following development directions of carbon fiber composite core aluminum conductors are proposed: Select fiber materials of low price, lightweight, and high strength; but after the formation of conductor only coefficient of linear expansion close to the conventional conductor and used to replace all or part of high-performance carbon fiber to form the price of conventional conductor close to the composite core conductor, which is widely used in various voltage levels including ultrahigh voltage (UHV) and UHV to maximize the comprehensive technical advantages of carbon fiber composite core aluminum conductors.

7.6 SUMMARY

This chapter analyzes and demonstrates the application of carbon fiber composite core aluminum conductor mode and scope of application, including line capacity increase and transformation, and the application of large span lines, widely used in various voltage lines to generally improve the safe operation of the line margin and generally reduce the line loss. At the same time, further research and development of low-cost carbon fiber composite core aluminum conductor should focus on breaking the bottleneck in terms of its application, and using a wide range of carbon fiber composite core aluminum conductor materials from the initial design stage of infrastructure to maximize its advantages.

REFERENCE

[1] IEC 62217 Ed.1. Polymeric insulators for outdoor and outdoor use with a nominal voltage greater than 1000V-General definitions, test methods and acceptance criteria, 2004.

CHAPTER 8

Application of Carbon Fiber Composite Core Aluminum Conductor

Contents

8.1 THE APPLICATION OF FOREIGN PRODUCTS

Carbon fiber composite core aluminum conductor is made in developed countries such as the United States, and was applied as a test line in aslett, Texas as early as August 2004. Table 8.1 shows the statistics of some overseas applications of carbon fiber composite core (CFCC) aluminum conductors. There are 13 in the United States, of which one is a test line with a length of 1.06 km. The number of 230 kV lines is 2, 4.0 km in total; the number of 161 kV line is 1, length is 0.15 km; the number of 138 kV lines is 3, a total of 34.96 km; the number of 115 kV line is 1, length is 0.14 km; and the number of 69 kV lines is 2, 1.37 km in total. The number of 34.5 kV lines is 2, 35.05 km in total and the number of 13.8 kV lines is 1, length is 1 km.

In 2006, Jiangsu Far East Holding Group Co., Ltd. and US CTC cofounded the Far East Composite Technology Co., Ltd. (Far East Company), specializing in the production of CFCC aluminum conductor. Early on, the CFCC aluminum conductor used in domestic projects was directly imported from US CTC by Far East. Far East still imports CFCC,

Engineering Energy Aluminum Conductor Composite Core (ACCC) and its Application
DOI: https://doi.org/10.1016/B978-0-12-815611-7.00008-9

Table 8.1 Overseas application of carbon fiber composite core aluminum conductor list [1]

S/N	Using location/country	Voltage class (kV)	Model and specification	Circuit length (km)	Number of circuit	Installation time
1	EPRI Haslett, TX/USA	Testing circuit	Drake	1.06	Single circuit	August 2004
2	National Grid-Niagara Falls, NY/USA	230	Hawk	1.20	Single circuit	January 2005
3	Austin Energy-Austin, TX/USA	138	Drake	0.20	Single circuit	March 2005
4	APS-Phoenix, AZ/USA	69	Drake	0.37	Single circuit	March 2005
5	City of Holland Holland, MI/USA	69	Hawk	1.00	Single circuit	April 2005
6	Xcel Energy, Denver, CO/USA	115	Hawk	0.14	Single circuit	May 2005
7	City of Kingman-Kingman/KS	34.5	Hawk	0.20	Single circuit	August 2005
8	EDF-Renardieres/France	Testing circuit	Drake	0.80	Single circuit	August 2005
9	Kingman (FEMA-Kingman), KS/USA	13.8	Hawk	1.00	Single circuit	September 2005
10	PacifiCorp-Salt Lake City, UT/USA	138	Drake	11.00	Single circuit	November 2005
11	WAPA Phoenix, AZ/USA	230	Drake	2.80	Single circuit	December 2005
12	AEP San Antonio, TX/USA	138	Drake	23.76	Single circuit	January 2006
13	City of Kingman-Kingman, KS/USA	34.5	Drake	34.85	Single circuit	July 2006
14	RTE-EDF-Minerve/France	Testing circuit	Drake	0.40	Single circuit	August 2006
15	AEP-Roaers, AR/USA	161	Drake	0.15	Single circuit	December 2006

aluminum strand to China. The core technology of composite core raw materials (carbon fiber, resin matrix) and formulation, process, etc. are foreign-owned, and do not have independent intellectual property rights. Given its background in foreign cooperation, its products will be positioned in foreign technology products.

At present, Far East Company CFCC aluminum conductor applications in the domestic performance is the most, since June 2006 in Fujian Longyan and Liaoning Liaoyang 220 kV line for the first time since the network operation. As of March 2013 Far East CFCC aluminum conductor has been used in 173 power transmission and transformation projects in 26 provinces and cities. The voltage level is 10–330 kV and the conductor length is 4030 km. It is mainly used for line capacity increase and transformation and some application projects are shown in Table 8.2.

Domestic manufacturers using foreign technology to produce CFCC aluminum conductor are Wuxi Huaneng Cable Co., Ltd. and TBEA Co., Ltd. Xinjiang Cable Factory (TBEA Xinjiang cable factory), but its application is rare. Among them, Wuxi Huaneng Cable Co., Ltd. started in 2008 to develop CFCC aluminum conductor. CFCC aluminum conductor is imported from Mercury in the United States, and TBEA Xinjiang Cable Factory purchased multistrand stranded CFCC from Tokyo Shibu Co., Ltd., Japan for the production of aluminum conductor carbon fiber reinforced (ACFR). The situation of the products of these two manufacturers is similar to that of the Far East Company. The core technologies are all held by the foreign parties.

8.2 APPLICATION OF SELF-DEVELOPED NEW TECHNOLOGY

The key science and technology project "Carbon fiber composite core aluminum conductor of independent research and development and application" managed by state grid corporation of china (SGCC) focuses on scientific research, product manufacturing, design, construction, and operation and maintenance of electricity and many other units of technical strength for the first time to achieve CFCC aluminum conductor from the basic materials and manufacturing processes until research and development on the application of this technology is completely independent. Based on the project achievements, it has formed a production base of CFCC aluminum conductors such as Handan Silicon Valley New Materials Co., Ltd. (Hebei Silicon Valley Chemical Co., Ltd.) and Zhongfu Carbon Core Cable Company. Among them, Handan Silicon

Table 8.2 Part of the application engineering for Far East Company carbon fiber composite core aluminum conductor [2]

S/N	Name of project	Project time (years)	Project site	Owner's unit
1	220 kV Yuen Tin Tong–Cao Xi line	2006	Longyan, Fujian province	Fujian Longyan Electric Power Bureau
2	110 kV Mengcun one land area substation line	2006	Wuxi, Jiangsu province	State Grid Wuxi Power Supply Company
3	110 kV Changge line	2006	Changzhou, Jiangsu province	Changzhou Power Supply Company
4	220 kV Qiancao line	2006	Liaoyang, Liaoning province	Liaoyang Power Company
5	220 kV Panxing line	2006	Panjin, Liaoning province	Panjin Power Supply Company
6	220 kV Jiaoli I loop wire	2006	Xiamen, Fujian province	Xiamen Electricity Bureau
7	220 kV Nanping Gutian–Yangzhen line	2007	Nanping, Fujian province	Xiamen Electricity Bureau
8	220 kV Hexi, Hexi double circuit	2007	Ningbo, Zhejiang province	Ningbo Power Company
9	220 kV Jingwei line	2008	Tianjin	State Grid Tianjin Electric Power Company
10	220 kV TianYuan line, Yuanwen line	2008	Tai'an, Shandong province	Taian Power Supply Company
11	330 kV Jishixia substation I, II line	2010	Xining, Qinghai province	Xining Power Company

Valley New Materials Co., Ltd. (Hebei Silicon Valley Chemical Co., Ltd.) uses domestic independent research and development technology in 2008 to build China's first annual production capacity of 200 t dry jet wet spinning high-performance carbon fiber precursor (T700 type) industrialized production lines. At present, it is expanding its annual output of 3200 t raw silk and 1200 t high-performance carbon fiber raw silk and supporting carbonation production line. The conductor CFCC production line has annual capacity of 20,000 km CFCC. Table 8.3 is the application of CFCC aluminum conductor produced by Handan Silicon Valley New Materials Co., Ltd. (Hebei Silicon Valley Chemical Co., Ltd.). From January 2009 to November 2013, a total of 120 projects were applied with a voltage level of 10–500 kV. The length of a single line is 2331 km, of which 500 kV Wanshun three-line has been in stable operation for 5 years. Table 8.4 shows the application of CFCC aluminum conductors produced by State Grid Liaoning Power Company. From May 2008 to December 2012, a total of seven projects were applied, with voltage levels of 66–500 kV and lead lengths of 141 km. In November 2009, the CFCC trapezoidal soft aluminum conductor produced by the new material of China Electric Power Research Institute was applied to the 220 kV Jinxian Line of the Zhejiang Jinhua Electric Power Bureau.

8.3 APPLICATION EXAMPLES

8.3.1 500 kV Application of EHV Lines

In January 2009, the domestically developed CFCC aluminum conductor was put into operation by hanging net in the 500 kV Wanshun three-line No. 470–473 Naizhang section, which was the first successful application of CFCC aluminum conductor in 500 kV extra high voltage (EHV) transmission line. LGJ-400/35 ACSR was replaced by JLNR/F2B-300/65 CFCC heat-resistant aluminum alloy twisted conductor, and the total length of the conductor, which has operated stably for 5 years, is 18 km.

8.3.1.1 Line Construction

1. Table 8.5 shows the parameter comparison of ACSR and CFCC aluminum conductor in Wanshun three-line. The data shows that although the tensile strength of CFCC aluminum conductor reaches 1.73 times that of ACSR, the mass per unit length reduces by 32%. Below the migration point temperature, the JLNR/F2B-300/65 has a coefficient of linear expansion of 60.5% of LGJ-400/35, whereas

Table 8.3 Example of self-developed carbon fiber composite core aluminum conductor applications

S/N	Name of project	Use unit	Installation place	Single circuit length (km)	Construction schedule
1	Beijing 500 kV Wanshun three-line	North China Grid Company Limited	Shunyi, Beijing	18	January 2009
2	Zhoukou 110 kV He–Xie line	Zhoukou Power Supply Corporation	Ankang, Henan province	2	July 2009
3	Leshan 110 kV Jiangfeng line	Leshan Electric Joint Stock Company	Jiajiang, Sichuan province	290.8	October 2009
4	State Power Jiujiang Power Plant 220 kV booster station Xunsha line I Xunsha line I interval	Jiangxi Electric Power Company	Jiujiang, Jiangxi province	0.4	December 2009
5	Langfang 220 kV Shao– Zhai line	North China Grid Company Limited	Sanhe, Hebei province	31.1	May 2010
6	Capacity Increase and Transformation of Zhuankou Substation in Wuhan–110 kV Xushenzhuo Line	Hubei Electric Power Company	Hanyang Development Zone, Wuhan, Hubei province	7.88	May 2010
7	Nanjing Meishan Iron and Steel 110 kV padpower lines 1 and 2	Meishan Iron and Steel Company	Yuhua District Nanzu Jiangsu	18.96	May 2010
8	Nanjing 220 kV Dongke–Yinli line	Nanjing Huaheng Transmission and Distribution Co., Ltd.	Nanjing, Jiangsu province	13.72	October 2010
9	Shanxi Xinzhou 35 kV Beizhao line	Shanxi Xinzhou Power Supply Bureau	Dai County, Xinzhou, Shanxi province	47.6	March 2011
10	Shanxi 220 kV Taoping line	Shanxi Electric Power Company	Xiezhou, Yuncheng, Shanxi province	18 (Dural)	March 2011
11	Shanxi Yangquan 110 kV Hongcan line	Shanxi Electric Power Company	Pingding County, Yangquan, Shanxi province	7.93	March 2011

12	Wuhan–110 kV Deao line	Wuhan Electric Power Company	Wuhan city, Hubei province	10.5	April 2011
13	Baolan iron 35 kV Shizuishan traction substation	Yinchuan Power Supply Bureau	Shizuishan city, Ningxia province	26.39	March–April 2012
14	110 kV Rantun power transmission I Foqin line	Henan Electric Power Company	Zhengzhou, Henan province	20.35	June–October 2012
15	220 kV Jiashan double circuit line	Hebei Grid Company Limited	Tangshan, Hebei province	56.92	–
16	110 kV Yulin Ding–Zhuan line	Yulin Power Supply Bureau	Dingbian, Yulin, Shaanxi province	75.39	–

Source: Handan Silicon Valley New Materials Co., Ltd.

Table 8.4 Self-developed carbon fiber composite core aluminum conductor applications

S/N	Name of project	Use unit	Single circuit length (km)	Delivery date
1	66 kV Wentao transmission line renovation project	Shenyang Power Company	27.60	May 2008
2	66 kV Changqing Jiayi power transmission line transformation project	Shenyang Power Company	7.38	November 2008
3	220 kV Niuhai line reconstruction project	Anshan Power Supply Company	22.16	September 2008
4	66 kV Chengfan transmission line renovation project	Shenyang Power Company	22.23	July 2009
5	220 kV Chaolong line 1 transmission line reconstruction project	Chaoyang Power Company	23.19	November 2009
6	500 kV Suizhong Power Plant to Gaoling line 3 new construction	Huludao Power Company	18.36	May 2010
7	66 kV Panjin Shuangshan second-line reconstruction project	Panjin Power Supply Company	20.50	December 2011

Source: Shenyang Cable Co., Ltd.

above the migration point temperature it drops to 7.8%. The conductor mass and the coefficient of linear expansion determine that the CFCC aluminum conductor has a significant sag advantage.

2. Due to the ductile and flexible characteristics of the soft aluminum and the weak friction between the composite core and the aluminum strands, the CFCC soft aluminum twisted conductor can easily lead to the shrinking core during the process of releasing and tightening the conductor, resulting in complicated construction such as follows: special shrink-proof fittings must be installed at the end of the conductor when soft aluminum twisted conductor tension stringing; the gripper for conventional conductor should not be used when tightening the line; and a preformed armor rod strain clamp that is laborious to install needs to be used. CFCC aluminum conductor aluminum strand used in Wanshun three-line is heat-resistant aluminum alloy, whose hardness and mechanical strength are higher than soft aluminum. To explore the construction performance of the conductor and determine

Table 8.5 Conductor parameter of 500 kV Wanshun three-line

Conductor model	LGJ-400/35	JLNR/F2B-300/65
Aluminum strand number/diameter per strand (piece/mm)	48/3.22	26/3.83
Number of steel (carbon fiber) strands/diameter per strand (piece/mm)	7/2.50	1/8.95
Aluminum cross section/core cross section (mm^2)	390.88/34.36	299.6/62.9
Comprehensive cross section (mm^2)	425.24	362.5
Outer diameter (mm)	26.82	24.17
Unit quality (kg/km)	1.349	0.920 35
Calculated tensile force (kN)	103.9	180
Elastic modulus (kN/mm^2)	65	71.45
Coefficient of linear expansion ($°C^{-1}$)	20.5×10^{-6}	12.4×10^{-6} (below 110°C) 1.6×10^{-6} (over 110°C)

Figure 8.1 Comparison test of construction fittings and machinery.

the construction fittings and machinery, test the traction effect of snake leather, gripper, and preformed armor rod strain clamp on CFCC heat-resistant aluminum alloy twisted conductor before construction, as shown in Fig. 8.1. The test indicated that the three kinds of construction fittings and machines meet the traction requirements for the conductor during construction. In case of not using special antishrinkage core fittings, there is no shrinking core phenomenon of the tested conductor. On the basis of the test, it is determined that the tension stringing of the CFCC heat-resistant aluminum alloy twisted conductor can directly adopt a conventional snakeskin sheath without adopting additional antishrinkage core measures. The gripper for conventional conductor can be used directly when tightening the conductor. Therefore, the construction difficulty of the CFCC heat-resistant aluminum alloy twisted conductor is greatly reduced compared with the CFCC soft aluminum strand, which is closer to the conventional wire construction, as shown in Figs. 8.2–8.5.

3. Scribing method of CFCC aluminum conductor is as follows. As the project uses the wedge joint strain clamp that is matched with the CFCC aluminum conductor, the conductor scribing method is determined by the three parameters of the lengths extending from the inner cone of the steel anchor + inner cone + composite core, which is unlike the conventional crimping type strain clamp scribing method.

Figure 8.2 Traction field.

Figure 8.3 Tension field.

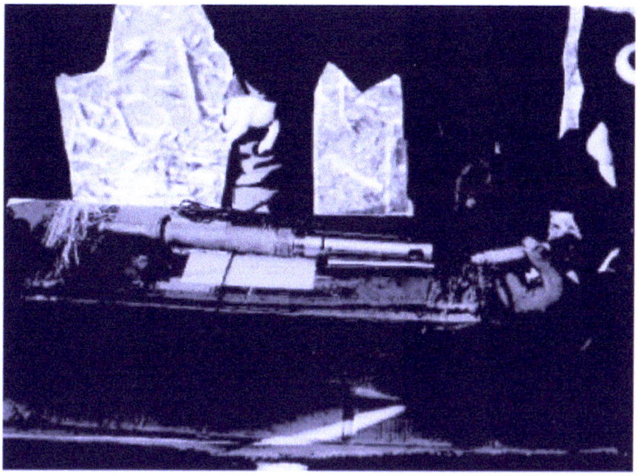

Figure 8.4 Conductor scribing method.

Figure 8.5 Check if the conductor is shrinking core and damaged.

4. For the four split and multisplit conductor, to avoid the drainage line and the adjacent subconductor being too close to each other, and thus touching and wearing each other out, strain clamp drainage plates of one phase conductor of the same group are different in the direction of the design. This is not a problem in single-conductor and double-split conductor construction of 110–220 kV line, and conventional conductor crimping type strain clamp design has long been used. However, the conventional strain clamps are different in the direction of 90 degrees from the wedge joint strain clamps. Therefore, in the design of extrahigh- and ultrahigh-voltage lines, the direction

adjustment of strain clamp drainage plates of the same group of one phase conductor should be taken to avoid mistakes.

5. Selection of strain tower drainage line (jumper): Capacity increase and transformation are performed with CFCC aluminum conductor, and strain tower drainage line should be replaced at the same time so that the flow capacity thereof increases correspondingly. If the same model of CFCC aluminum conductor replaces the original drainage line, due to the large drainage line bending angle and low CFCC winding performance, the installation is difficult and it is easy to break the drainage line, or even damage the drainage tube or drainage board, which is more prominent for the relatively short drainage lines below 110 lines. In addition, a "bamboo" phenomenon occurred when CFCC aluminum conductor was found by a certain unit in China to not be appropriate for use as a drainage line at high temperature. In view of the small tension of the conductor to be drained and the problem of arc sag, it is required that the drain wire be replaced with a thermal-resistant aluminum alloy conductor (or a steel core and soft aluminum strand) of the same size as the CFCC aluminum conductor to ensure the flow capacity of the drainage line, and to prevent mechanical hazards or defects of the drainage line.
6. Supporting capacity expansion of substation on both sides of the line: With CFCC aluminum conductor for line capacity increase and transformation, the relevant equipment and conductor capacity within substations at both sides of the line should be checked at the same time, including circuit breakers, disconnectors, current transformers, wave filters, leads, and so on.

8.3.1.2 Line Running

As of January 2014, the JLNR/F2B-300/65 CFCC aluminum conductor of the 500 kV Wanshun three-line has been operating continuously and steadily for 5 years without electrical and mechanical faults and without electrical and mechanical defects on the conductor. The parameters of ground electrical/magnetic field strength, audible noise, radio interference, tilt and twist of tower, conductor and fittings temperature, conductor sag, and distance to ground are all measured to meet the relevant standards.

1. Conductor, fittings temperature test

   On January 9, 2009 Wanshun three-line three returned to operation, the JLNR/F2B-300/65 CFCC aluminum conductor and fittings of 470–473, the LGJ-400/35 ACSR and fittings of 470 small side and 473 large side undergo infrared temperature measurement.

a. Conductor temperature test. At the beginning of operation (January 12, 13, 17), the CFCC aluminum conductor was replaced at about 1000 MW load, and the temperature was lower than the adjacent span ACSR by 5–10°C; later, the temperature difference gradually reduced until it was close. The reason for the difference in the temperature of the two conductors during the initial run is mainly due to the bright new surface of the new conductor and the dark color of the surface of the old conductor, as shown in Fig. 8.6. Conductor carrier flow calculation formula:

$$I_t = \left\{ \frac{9.92\theta(VD)^{0.485} + \pi\varepsilon SD[(\theta + t_a + 273)^4 - (t_a + 273)^4] - \alpha_s I_s D}{k_t R_{dt}} \right\}^{0.5}$$

where α_s is the conductor endothermic coefficient, coefficient of the bright new line is 0.23–0.46, and the dark old line is 0.90–0.95, that is, the new line of heat absorption is relatively small; ε is the emissivity of the conductor surface, the bright new line is 0.23–0.46, black old line is 0.90–0.95, that is, the new line's radiation heat is relatively small.

On one hand, the new line absorbs less heat, resulting in a lower temperature of the new line; on the other hand, temperature is measured by the infrared imager, which determines the temperature of a measured object mainly depending on heat radiation of the measured object. The new line (CFCC aluminum conductor) surface is bright, with small emissivity, less heat radiation, leading to

Figure 8.6 Bright carbon fiber composite core aluminum conductor.

the low-temperature measurement of the new line. But the former factor reflects the actual temperature of the conductor; the latter factor is only a matter of measurement principle and does not reflect the actual temperature of the conductor. If possible, contact measurement methods are recommended and the results should be more accurate.

b. Fittings temperature test. During the same period of time, there is no difference (less than 3°C) in the temperature of the strain clamps and the strain clamp drainage lines of the two kinds of conductors; that is, the wedge joint strain clamps matched with the CFCC aluminum conductor operate normally.

2. Conductor sag and distance-to-ground test

After Wanshun three-line resumed operation, the 471−472 CFCC aluminum conductor and 465−466 ACSR arc sag were measured for comparison.

a. Conductor sag measurement. Since January 9, 2009, the Wanshun three-line phase conductor, estimated to be completed in January 2017, 471−472 span CFCC aluminum conductor arc sag has had a nearly 300 mm increase in volume, which is mainly due to conductor temperature rise and conductor creep after the line was put into operation; the load of the Wanshun three-line is about 1000 MW.

As of January 2, 2010, the CFCC aluminum conductor curve sag has tended to stabilize, as shown in Fig. 8.7. After replacing the CFCC aluminum conductor, the arc sag is smaller than the same

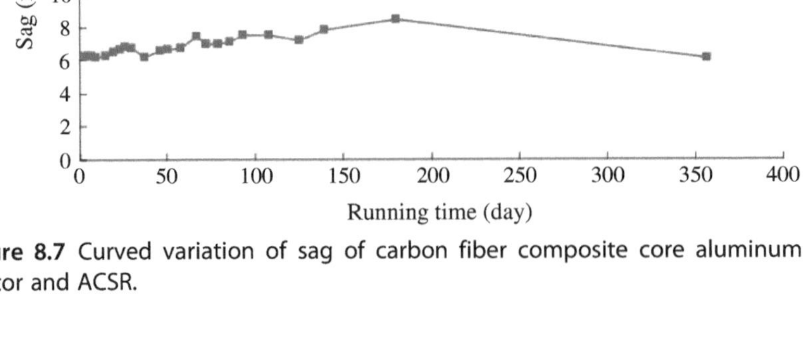

Figure 8.7 Curved variation of sag of carbon fiber composite core aluminum conductor and ACSR.

span (471–472) the original ACSR by 3–4 m. Compared with the measured ACSR Wanshun three-line No. 465–466 span ACSR, the sag effect is more significant.

Creep characteristics of the conductor in the initial operation of the line include a relatively rapid plastic elongation, leading to arc sag increase; later, elongation tends to stabilize.

b. Conductor clearance to ground. After Wanshun three-line resumed operation, the 471–472 CFCC aluminum conductor and 465–466 ACSR distance to ground was measured, the distance to ground curve is shown in Fig. 8.8. As of January 2, 2010, CFCC aluminum conductor and ACSR had no significant changes in distance to the ground, and the CFCC aluminum conductor has entered relatively stable operation.

3. Ground field, radio interference, noise test

a. Electric field, magnetic field. The test results show that field strength (including electric field, magnetic field) of the span from the CFCC aluminum conductor is less than that of the conventional conductor. The test results are mainly due to the mass advantage of the CFCC aluminum conductor leading to slack significantly less than the conventional conductor.

b. Radio interference. As stipulated in GB15707-1995 *HV AC Overhead Power Transmission Line Radio Interfering Limit Value*, at a distance of 20 m from the phase conductor projection, the radio

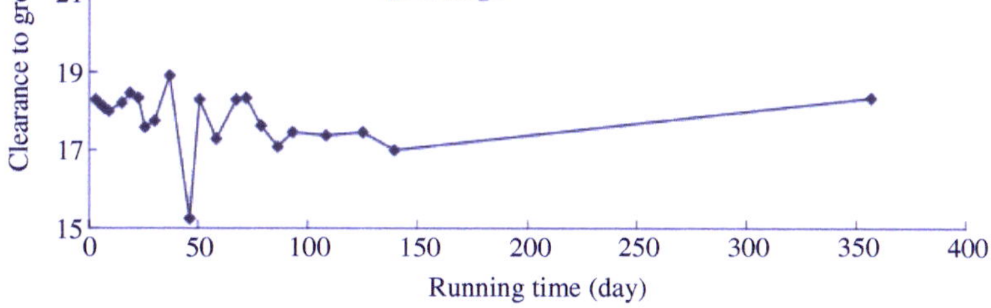

Figure 8.8 The distance to ground curve of sag of carbon fiber composite core aluminum conductor and ACSR.

interference limit of the 500 kV line is 55 dB at 0.5 MHz and 50 dB at 1 MHz. Wanshun three-line radio interference after replacement of CFCC aluminum conductor meets the requirements.

c. Audible noise. Wanshun three-line replaces CFCC aluminum conductor, and the noise value generated is slightly larger than the original conductor and the other two lines. The noise value immediately below the center conductor is the largest, with daytime and nighttime values of 39.3 and 34.7 dB, respectively, which are much lower than the daytime 55 dB and nighttime 45 dB noise standards stipulated in the GB3096-2008 *Acoustic Environment Quality Standard* constituting noise pollution.

The noise value at 60 m outside the conductor is not much different from the local background value, indicating that the line noise is basically unaffected after attenuation. The reason for the slightly larger measured value is the burr on the surface of the new conductor and the low corona voltage. After a certain period of operation, the noise value can be properly reduced.

8.3.2 Application of Continuous Tension Pipe With Continuous Tension Stringing Technology

After the successful development of crimping CFCC aluminum conductor supporting fittings (including straight connecting pipe, strain clamp), it performs many on-site tests and can be applied to actual projects.

1. In May and June 2009, the CFCC aluminum conductor and its matching crimp-type straight connecting pipe were successively tested for 500 kV Zhangmen Line and concierge second line to successfully achieve continuous tension stringing of CFCC aluminum conductor with straight connecting pipe. The conductor model JLNR/F2B-240/45, the middle is crimping of the five crimp tube. The trial run section is located on a mountain ridge, with many towers turning angles more than 30 degrees and even close to 60 degrees. 822 double tackles are used for all tackle, and 822 double tackle are used for towers with turning angle greater than 30 degrees. Straight connecting pipe outsourcing steel armor for protection, the CFCC aluminum conductor with a continuous tube passed all the tackle.

   After the pay-off, the conductor and crimping straight connecting pipe should be recovered and checked: Peel the aluminum strands to check the CFCC near the continuous tube for damage. It is tested on

Figure 8.9 110 kV Jiangfeng line capacity increase and transformation project.

the conductor and crimping straight connecting pipe for the tensile force, and compared with the sample without tension stringing, the tensile force has no significant decline.

2. In October 2009, in the capacity increase and transformation project Sichuan Leshan Electric Power Co., Ltd. 110 kV Jiangfeng line, once again successfully achieved CFCC aluminum conductor with crimping straight connecting pipe continuous tension stringing. The conductor model was JLNR/F2B-300/65. The line is located in a hilly area. The straight connecting pipe in the pay-off construction passes through two corner towers greater than 30 degrees (with angles of 41 and 36 degrees, respectively), to select 822 double tackle. Other towers with angle less than 30 degrees use 822 single tackle. Straight connecting pipe outsourcing steel armor for protection (see Fig. 8.9), the CFCC aluminum conductor with a continuous tube passed all the tackle. As of December 2013, the CFCC aluminum conductor and crimping-type straight connecting pipe have been running stably for 4 years and 2 months.

8.4 SUMMARY

This chapter summarizes the development and applications of CFCC aluminum conductor both at home and abroad. Among them, the independent research and application of carbon fiber composite core aluminum conductor, a key technology project managed by SGCC, has for the first time realized the complete domestic independent R&D of CFCC aluminum conductor from basic materials and manufacturing technology to application technology. The formation of annual production capacity of 20,000 km of CFCC production base, wire, fittings and construction, operation, and maintenance technology has been applied to 110 (66)–500 kV overhead transmission lines, breaking the monopoly of developed countries on materials and technology, and laying a solid foundation for the widespread popularization and application of CFCC aluminum conductors. This is of strategic significance for construction and upgrading of power grids, conservation of land resources, and promotion of the country's high-performance carbon fiber industry.

REFERENCES

[1] Xuming L, Jun Y, et al. Review of new composite material core conductor technology. Power Syst Technol 2006;30(19):18–22.
[2] IEC 62217 Ed.1: Polymeric insulators for outdoor and outdoor use with a nominal voltage greater than 1000V-General definitions, test methods and acceptance criteria, 2004.

FURTHER READING

Chuanyong Y. Development of overhead transmission line composite conductor. Electric Power Constr 2004;11:25.
Diansheng Z. Power engineering high voltage transmission line design manual. Beijing: China Electric Power Press; 2003.
Houlgate RG, Swift DA, Cimador A, Pourbaix F, Marrone G, Nicolini P. Field experience and laboratory research on composite insulators for overhead line. France: CIGRE; 1986, paper 12–15.
IEC 61109, Composite insulators for a.c. overhead lines with a nominal voltage greater than 1000V-Definitions, test methods and acceptance criteria; 1992.

INDEX

Note: Page numbers followed by "*f*" and "*t*" refer to figures and tables, respectively.

D

E

J

L

M

N

O

P

Q

R

S

T

U

MIX
Papier | Fördert gute Waldnutzung
FSC® C083411

Kolibri 360 GmbH
Ferdinand-Jühlke-Straße 7
99095 Erfurt, Deutschland
produktsicherheit@kolibri360.de